LAB WORKBOOK

SECOND EDITION

Natural Resources Systems

Kasee Smith • Tara Berescik Burke

Publisher
The Goodheart-Willcox Company, Inc.
Tinley Park, IL
www.g-w.com

ISBN 979-8-89118-220-2

1 2 3 4 5 6 7 8 9 – 25 – 28 27 26 25 24 23

Image Credits. Front cover: digidreamgrafix/Shutterstock.com

Introduction

This lab workbook is intended to be used with the *Natural Resources Systems* textbook. The chapters in the lab workbook correspond to those in the textbook and should be completed after reading the appropriate textbook chapter. This lab workbook will help you understand and use methods of inquiry used in environmental sciences.

Answering questions in the lab workbook will help ensure that you have mastered the technical knowledge presented in the textbook. Several types of questions are used, including matching, true/false, identification, multiple choice, short answer, and essay questions. There are also independent and classroom activities, as well as various lab activities to perform in class. Some questions involve math calculations, and others may require the use of tables and charts. The questions are designed to test your technical knowledge of techniques covered in the textbook.

Kasee Smith

Tara Berescik Burke

Contents

Chapter 1

Introduction to Natural Resources 1

Activity 1A Agricultural Land Use 7

Chapter 2

Career Planning 9

Activity 2A Personal Leadership Plan 17

Activity 2B Speech Outline Worksheet......... 23

Activity 2C Career Inventory 27

Activity 2D Leaders in History................ 31

Chapter 3

Historical Perspectives of Natural Resources and Conservation 35

Activity 3A People Who Make a Difference... 41

Activity 3B Conservation Time Line........... 43

Chapter 4

Ecology and Earth 47

Activity 4A Biogeochemical Cycles 55

Chapter 5

Sustainability in the Environment 57

Activity 5A Energy Usage Lab 63

Activity 5B Biofuel Lab....................... 67

Activity 5C Carbon Footprints................ 69

Activity 5D The Organic Debate 71

Chapter 6

Habitat Destruction 75

Activity 6A Xeriscape Design Plan............ 79

Activity 6B Carrying Capacity 81

Chapter 7

Global Environmental Issues 85

Activity 7A Eating Local...................... 89

Activity 7B Foods around the Globe.......... 91

Chapter 8

Population Dynamics................ 93

Activity 8A Population Dynamics............. 99

Activity 8B Populations and Survival Needs 105

Activity 8C World Population Growth........ 109

Chapter 9

What Is Soil? 113

Activity 9A Edible Soil Profile................ 121

Activity 9B Comparing Soil Particles......... 125

Chapter 10

Soil Erosion, Leaching, and Pollution.......................... 127

Activity 10A Muffin Mining Lab 132

Activity 10A Soil Erosion 133

Chapter 11

Controlling Soil Erosion......... 137

Activity 11A Find the Science 141

Activity 11B Designing Scientific Experiments........................... 143

Activity 11C Communicating Results........ 147

Chapter 12

Mining of Natural Resources 149

Activity 12A Minerals in Everyday Objects... 153

Activity 12B Mineral Scavenger Hunt........ 155

Chapter 13

Water Supply........................... 159

Activity 13A The Water Cycle................ 165

Activity 13B Taste Testing Drinking Water.... 167

Activity 13C Water Olympics 169

Chapter 14

Water Quality 173

Activity 14A Testing Water Quality 179

Chapter 15

Wetlands 181

Activity 15A Groundwater Infiltration Lab ... 187

Chapter 16

Weather and Climate 191

Activity 16A Weather Forecast 195

Activity 16B Making Lightning 197

Activity 16C Lightning Lab Variations........ 199

Activity 16D Tornado in a Bottle............. 201

Activity 16E Tornado in a Bottle Variations... 203

Activity 16F Keeping a Weather Journal 205

Chapter 17

Air Quality 207

Activity 17A Technology Highlight 213

Activity 17B International Airsheds.......... 215

Chapter 18

Interactions 219

Activity 18A Food Chains and Food Webs................................. 225

Activity 18B Food Webs..................... 227

Chapter 19

Endangered Species 229

Activity 19A Persuasive Speech Outline 235

Activity 19B Captive Breeding Research Assignment........................... 240

Chapter 20

Fisheries 241

Activity 20A Environmental Science: Tragedy of the Commons (Part 1)....... 247

Activity 20A Environmental Science: Tragedy of the Commons (Part 2)....... 253

Chapter 21

Game Species 255

Activity 21A Casting Animal Tracks.......... 261

Activity 21B Limiting Factors................ 265

Chapter 22

Forests 269

Activity 22A Reading Tree Cuttings.......... 277

Activity 22B Calculating Board Feet 279

Chapter 23

Forest Succession and Management 281

Activity 23A Flower Dissection and Analysis 287

Activity 23B Forest in a Jar 289

Chapter 24

Threats to Forests 291

Activity 24A Fire Safety Messages 297

Chapter 25

Grasslands and Rangelands ... 299

Activity 25A Parts of a Flower 305

Activity 25B Parts of a Stem 307

Activity 25C Bureau of Land Management Wild Horses and Burros Control Method Debate 309

Activity 25D Beef Cattle on the Range 311

Chapter 26

Outdoor Recreation 313

Activity 26A Safety Pamphlet 317

Chapter 27

Natural Protected Areas 319

Chapter 28

Surveying and Cartography 323

Activity 28A Measuring Worksheet 327

Name ______________________ Date ______________ Period ____________

CHAPTER 1

Introduction to Natural Resources

Chapter Review

Carefully study the chapter and then answer the following questions.

1. _____ Which of the following are considered natural resources?
   A. Crude oil
   B. Water, air, and soil
   C. Plants and animals
   D. All of the above.

2. _____ Which of the following resources are plentiful and replenished naturally in a short period?
   A. The sun and air
   B. Crude oil and natural gas
   C. Precious metals
   D. All of the above.

3. _____ Living natural resources are referred to as _____.
   A. tropic
   B. abiotic
   C. geotropic
   D. biotic

4. _____ Nonliving natural resources are referred to as _____.
   A. biotic
   B. tropic
   C. abiotic
   D. geotropic

5. _____ Which of the following forms of energy is the cleanest form of energy we use?
   A. Wind energy
   B. Biomass energy
   C. Solar energy
   D. Geothermal energy

6. _____ *True or False?* The turbines on windfarms convert the kinetic energy of air in motion into electricity or mechanical power.

7. _____ *True or False?* All geothermal power plants are considered green.

8. _____ Which of the following statements is true?
   A. Biomass fuels are considered green power sources.
   B. Biomass energy is harnessed from decomposing organic materials.
   C. Biofuels are solid fuels made from organic materials.
   D. All of the above.

9. _____ *True or False?* Hydroelectric power plants are considered green power plants and have little effect on local environments.

10. _____ Which of the following statements about tidal energy is true?
   A. Tidal power generation may negatively affect marine species.
   B. Tidal power generation is clean and considered green power production.
   C. Tidal power is generated through the rise and fall of the tides.
   D. All of the above.

11. _____ *True or False?* Timber is considered a nonrenewable resource because it takes as many as 50 years for a tree to reach a harvestable size.

12. _____ Which of the following are nonrenewable resources?
   A. Timber, water, and air
   B. All abiotic resources
   C. Fossil fuels, soil, and minerals
   D. All of the above.

13. _____ Which of the following practices reduce soil erosion?
   A. Planting grassed waterways
   B. Maintaining vegetative plant cover
   C. Conservation tillage methods
   D. All of the above.

14. _____ *True or False?* Minerals such as gold, uranium, and potassium are considered renewable resources because they can be created through synthetic means.

15. _____ Nuclear energy is considered a nonrenewable resource because it _____.
   A. requires regulated disposal
   B. uses excessive amounts of water for cooling
   C. requires uranium
   D. All of the above.

16. _____ *True or False?* Fossil fuels are considered nonrenewable resources because they require millions of years to form.

Name ____

17. _____ Which of the following are abiotic resources?
   A. Bacteria and fungi
   B. Fossil fuels
   C. Wild animals
   D. Green plants

18. _____ *True or False?* Living plants and animals depend on nonliving resources to survive.

19. _____ Which of the following are fauna?
   A. Minerals
   B. Plants
   C. Sea kelp
   D. Deer

20. _____ Which of the following are flora?
   A. Gold and silver
   B. Maple trees
   C. Earthworms
   D. All of the above.

21. _____ Which of the following statements about biodiversity is true?
   A. It fosters healthy ecosystems.
   B. Flora and fauna contribute to biodiversity in ecosystems.
   C. It is the variety of organisms living in an ecosystem.
   D. All of the above.

22. _____ *True or False?* Environmental stewardship includes practices such as recycling and installing soil conservation structures.

23. _____ *True or False?* Conservation is the nonuse of natural resources.

24. _____ *True or False?* Preservation is the protection and nonuse of natural resources.

25. _____ We can all practice conservation with our actions, such as _____.
   A. using a compost bin for food scraps and yard waste
   B. repairing leaky faucets and watering landscaping with collected rainwater
   C. not running the water when brushing our teeth
   D. All of the above.

Vocabulary Review

Match each of the following terms with the appropriate definition or description.

A. abiotic resource
B. biomass
C. biotic resource
D. conservationist
E. environmental stewardship
F. fauna
G. flora
H. fossil fuels
I. natural resource
J. nonrenewable resource
K. organism
L. preservationist
M. renewable resource
N. solar energy
O. wind energy

1. _____ A person who advocates for the nonuse of natural resources.
2. _____ A person whose career focuses on educating others and implementing programs to conserve resources.
3. _____ A resource that cannot be replenished or requires a long time to be replenished.
4. _____ A resource that is a living organism.
5. _____ individual form of life, such as an animal, plant, or bacterium.
6. _____ A resource that is plentiful and replenished naturally in a relatively short period.
7. _____ A term that refers to animal life.
8. _____ Buried combustible deposits of crude oil, coal, and natural gas.
9. _____ The responsible use of natural resources and the protection of those resources that are in short supply.
10. _____ A resource that is nonliving.
11. _____ One of the many things found in our environment that we depend on for life.
12. _____ A term that refers to plant life.
13. _____ The kinetic energy of air in motion that can be converted into mechanical power or electricity.
14. _____ The organic material used to create fuel and power.
15. _____ The radiant energy of the sun that is used to create solar power.

Name ____________________

Thinking Critically

1. Is it possible to know in advance how our use of a resource will affect the environment? Explain your answer.

2. In terms of input and useful output, what does it mean when a manufacturer states that an appliance is 40% efficient? How efficient are the appliances in your home?

3. Why is environmental stewardship important to the sustainability of our natural resources?

Notes

Name ______________________ Date ______________ Period __________

Activity 1A

Agricultural Land Use

Objectives

- Identify the amount of land used to grow and raise food.
- Explain why sustainable agriculture is important to our food supply and the environment.

Materials

- activity worksheet
- markers, color pencils, or crayons

Global Land Use

The surface of Earth is about 197 million square miles. In order to understand the scope of agriculture, it is important to recognize the amount of land being used for agriculture and food production. Use the boxes on the completed grid to guide your analysis of agricultural land use.

Instructions

Agriculture has a large scope in relation to the amount of land it impacts. Read the passage and then color the items as directed.

1. It is estimated that 71% of Earth's surface is covered by water. To represent this area, color 71 of the boxes blue.
2. The remaining 29 boxes represent the land on Earth. Of this land, just over half is unsuitable for humans to live, graze animals, or grow crops. This includes areas in the arctic, deserts, the tops of rocky mountains, or swamps. Color 15 boxes brown to represent this area on Earth's surface.
3. The remaining 14 boxes represent the areas on Earth which are habitable, or suitable for human life. Of this land, about 42% is covered by towns, cities, and buildings. Color six boxes red to represent this portion of the land.
4. The remaining nine boxes are available for grazing, pastures, and growing crops. About 55% of this land is available for grazing, but not suitable for growing crops. Color five boxes yellow to represent this land area.

5. The last three boxes represent the arable cropland currently in use for agricultural production in the world. Color them green.

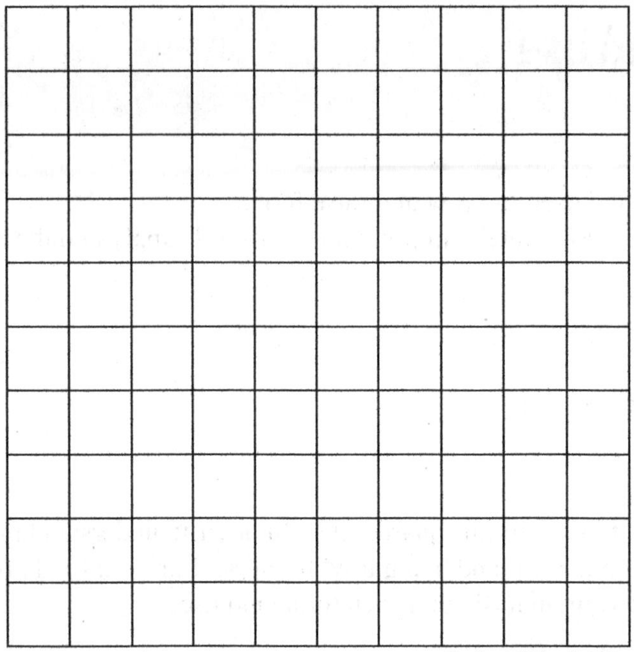

Goodheart-Willcox Publisher

6. What surprises you about the graphic you just completed? Did you expect the graphic to show only three green boxes?

7. Some experts estimate that cropland in the United States is being lost at the rate of up to 3000 acres a day. How would you share with the public the importance of preserving cropland?

Name ______________________________ Date ______________ Period ____________

CHAPTER 2 Career Planning

Chapter Review

Carefully study the chapter and then answer the following questions.

1. _____ Paid employment that involves handling one or more jobs is referred to as a(n) _____.
   A. career
   B. task
   C. occupation
   D. profession

2. _____ A series of related occupations that show progression in a field of work is a(n) _____.
   A. career
   B. task
   C. occupation
   D. profession

3. _____ *True or False?* The purpose of a career cluster is to prepare students to transition from school to a rewarding career.

4. _____ Knowing the relationship of different careers in a career pathway is helpful because _____.
   A. it allows you more flexibility when searching for employment
   B. the skills for different jobs in a similar field may overlap
   C. you will be able to adapt to changes in your occupation
   D. All of the above.

5. _____ *True or False?* Before setting career goals, you should consider your values, interests, abilities, and aptitudes.

6. _____ It is important to examine your likes and dislikes to _____.
   A. determine your abilities and aptitudes
   B. ensure you earn a high salary
   C. identify the type of tasks you would enjoy in a career
   D. All of the above.

7. _____ The skills you develop with practice are _____.

A. unimportant
B. abilities
C. aptitudes
D. All of the above.

8. _____ The ability to learn something quickly and easily is a person's _____.

A. aptitude
B. ability
C. attitude
D. personality

9. _____ *True or False?* A person's personality traits should have no influence on a person's chosen occupation.

10. _____ *True or False?* You need to determine the level of education that is required and your willingness to pursue such an education.

11. _____ An arrangement with a company or educational institution whereby a student is supervised by an experienced person is a(n) _____.

A. apprenticeship
B. internship
C. job shadow
D. management position

12. _____ The process of identifying something you want to accomplish and establishing a plan to achieve the desired result is referred to as _____.

A. task setting
B. aptitude setting
C. goal setting
D. All of the above.

13. _____ Actions that a person wants to accomplish over a span of several years are _____.

A. long-term goals
B. short-term goals
C. action goals
D. planning goals

14. _____ Actions that a person wants to accomplish in the near future are _____.

A. long-term goals
B. short-term goals
C. a plan of action
D. planning goals

15. _____ A list of individual steps that a person will take to achieve a goal is a(n) _____.

A. goal plan
B. action chart
C. plan of action
D. short list

16. _____ SMART goals are _____.

A. a plan of action

B. short, metered, available, ready, and technical

C. scholarly, misguided, achievable, realistic, and tentative

D. specific, measurable, attainable, realistic, and timely

17. _____ Goals that are unimportant or have little value to you will *not* be _____.

A. achievable

B. motivational

C. reliable

D. tempting

18. _____ Which of the following are useful when you are job searching?

A. Social networking sites

B. Networking

C. Online postings

D. All of the above.

19. _____ *True or False?* Joining organizations can help you learn about career options and network with others who may help you establish a career.

20. _____ A brief outline of your education, work experience, and other qualifications for work is a _____.

A. résumé

B. portfolio

C. cover letter

D. reference

21. _____ *True or False?* It is not necessary to obtain permission from an individual to list them as a reference because the potential employer will not contact him or her.

22. _____ *True or False?* Employers may request that an e-portfolio be included with your résumé.

23. _____ Which of the following should be included in a cover letter?

A. Where you heard about the job

B. Title of the job you seek

C. When you are available to begin work

D. All of the above.

24. _____ When preparing for a job interview, you should _____.

A. decide what to wear

B. research the employer

C. practice interviewing

D. All of the above.

25. _____ Interpersonal skills include activities, such as _____.

A. dressing to impress your coworkers
B. teaching others and leading
C. showing work shortcuts to your coworkers
D. All of the above.

26. _____ *True or False?* Good leaders encourage members of a group to focus on their individual goals before focusing on the group's goals.

27. _____ *True or False?* Placing the desires and welfare of others above yourself is a good leadership trait.

28. _____ *True or False?* Trying to understand the ideas of others before trying to get others to understand your ideas is an effective skill for team members.

29. _____ *True or False?* Good personal hygiene is not a factor that employees consider when interviewing potential employees.

30. _____ Which of the following are time-wasting behaviors?

A. Updating your social media during work hours
B. Visiting with coworkers to socialize
C. Making personal calls and texting during work hours
D. All of the above.

31. _____ *True or False?* Your outlook on life can often determine the success you have on your job.

32. _____ *True or False?* Keeping your work area well-organized helps you perform tasks more efficiently.

33. _____ *True or False?* Behaving professionally includes responding appropriately to constructive criticism.

34. _____ Verbal communication involves _____.

A. body language
B. speaking, listening, and writing
C. facial expressions
D. All of the above.

35. _____ Nonverbal communication involves _____.

A. speaking, listening, and writing
B. arguing and negotiation
C. body posture and facial expressions
D. All of the above.

36. _____ Cell phone etiquette includes _____.

A. speaking quietly in public places
B. not answering during a meeting
C. turning off your phone during meals
D. All of the above.

Name ____

37. _____ Which of the following is *not* professional email etiquette?
   A. Marking every message as being urgent
   B. Using all capital letters and emoticons
   C. Writing extensively detailed messages
   D. All of the above.

38. _____ *True or False?* Ethical behavior on the job applies to words, actions, computer use, and the use of social media.

39. _____ *True or False?* The material you post on your personal media accounts *cannot* affect your employability.

40. _____ *True or False?* Employers can exclude candidates for certain employment positions based on the applicant's religion or nationality.

41. _____ *True or False?* It is a good practice to tell your employer how much you hated working for him or her when you are leaving for a new job.

Vocabulary Review

Match the following terms with the appropriate definition or description.

A. abilities
B. apprenticeship
C. aptitude
D. associate's degree
E. bachelor's degree
F. career
G. conflict
H. cover letter
I. diversity
J. entrepreneur
K. ethical behavior
L. leadership
M. long-term goal
N. master's degree
O. negotiation
P. plan of action
Q. portfolio
R. punctual
S. self-motivation
T. short-term goal

1. _____ A collection of items that show your special achievements and accomplishments.
2. _____ Skills that a person develops with practice.
3. _____ Differences among people regarding gender, sexuality, religion, race, nationality, and many other distinctions.
4. _____ A college degree that requires another year or two of study beyond a bachelor's degree.
5. _____ The ability to guide and motivate others to complete tasks or achieve goals.
6. _____ A document sent with a résumé to provide information on why you are qualified for the position for which you are applying.
7. _____ A hostile situation resulting from opposing views.
8. _____ The process of working toward an agreement for which all parties give and take.
9. _____ A two-year college degree.

10. _____ An ability to learn something quickly and easily.

11. _____ A person who runs his or her own business and works for himself or herself.

12. _____ An arrangement in which a person learns a trade under the direction and guidance of an expert.

13. _____ A college degree usually requiring four years of study.

14. _____ An inner urge to perform well.

15. _____ An item or action that you wish to accomplish in the near future, usually less than 12 months.

16. _____ A series of related occupations that show progression in a field of work.

17. _____ An item or action that you wish to accomplish over the span of several years or even a lifetime.

18. _____ Conforming to accepted standards of fairness and good conduct.

19. _____ A list of individual steps that you will take to achieve a goal.

20. _____ Prompt and on time.

Thinking Critically

1. How do we know how to make good decisions?

Name ____________________

2. How is conflict an inevitable part of relationships?

3. How do we respect and value diversity?

Notes

Name ______________________________ Date ______________ Period __________

Activity 2A

Personal Leadership Plan

Objectives

- Evaluate your leadership skills and style.
- Create a personal guide for your career goals.
- Establish a time line in which you will achieve your goals.

Personal Inventory

Who are you as a leader? This worksheet is designed to help you evaluate your own leadership style and create a guiding direction for a specific goal. Answer the following questions.

1. What are your greatest strengths?

____

____

____

____

____

2. What have you done in your life that you take the most pride?

____

____

____

3. Who is the person you look up to most in the world?

____

____

4. List five characteristics of the person above that you would like to have (or have more of).

____

____

____

____

____

5. If you could change one thing about the world, what would it be?

____

____

____

____

6. If you could change one decision you have made in your life, what would it be?

7. If you could try one thing in the next year knowing for certain you would not fail, what would it be?

Setting Goals

Now that you have examined a little bit about what might guide you to be a leader, select a goal you would like to work toward over the course of your high school experience.

8. What is your goal? Be certain to write a goal that is SMART.

9. What characteristics do you already possess that will help you reach this goal?

Name ______________________________

10. Who in your life could help you with this goal? List their name and a description of how they could help.

A. Name: ______________________________

How they can help you achieve your goal: ______________________________

____

____

____

B. Name: ______________________________

How they can help you achieve your goal: ______________________________

____

____

____

C. Name: ______________________________

How they can help you achieve your goal: ______________________________

____

____

____

D. Name: ______________________________

How they can help you achieve your goal: ______________________________

____

____

____

11. What skills will you need to learn or practice in order to reach this goal?

____

____

____

____

12. Create a goal time line. Break your goal into 3–5 smaller parts or steps. Assign a due date to each of the smaller parts.

Goal: ______________________________

A. Step 1: ______________________________

____

____

Step 1 due date: ______________________________

B. Step 2: ____________________

____

____

Step 2 due date: ____________

C. Step 3: ____________________

____

____

Step 3 due date: ____________

D. Step 4: ____________________

____

____

Step 4 due date: ____________

E. Step 5: ____________________

____

____

Step 5 due date: ____________

13. What are the potential obstacles to you obtaining your goal?

____

____

____

____

____

____

____

____

14. How could you overcome the obstacles you listed?

____

____

____

____

____

____

____

____

Name ____

15. Why is this goal important to you?

____

____

____

____

____

____

____

____

16. Finally, take a separate piece of paper and write this goal down. Place the paper somewhere you will see it often.

Personal Assessments

If you want to know more about your own personal leadership style and preferences, there are specific assessments that will help you understand more about who you are. Ask your teacher about how to take personal inventories, such as the Myers-Briggs Type Indicator, the StrengthsFinder assessment, or Kolb's Learning Style Inventory.

Notes

Name ______________________ Date ____________ Period ________

Activity 2B

Speech Outline Worksheet

Objectives

- Gather and organize information in preparation for a speech.
- Develop organization and preparation skills.

Instructions

This form is designed to help you gather the information that you will be including in a speech. Carefully consider potential topics or use topics assigned by your teacher.

Choose a Topic

1. Speech Topic: ______________________
2. Speech Title: ______________________

Identify Your Audience

1. Who is your intended audience?

____

____

2. How much about your topic do you think your audience knows?

____

____

3. What do you want your audience to be able to take away from the speech? What is the thing you want them to remember?

____

____

4. What are the challenges you think you will have with this audience?

____

____

5. How will you overcome those challenges?

____

____

Section One: The Introduction

6. What are you going to use to catch the audience's attention? (Be specific and tell exactly what you are going to do for your attention grabber.)

____

____

____

____

Section Two: The Body

7. What are your main points? (Identify at least three.)

   A. Section 1: ____

   ____

   ____

   B. Section 2: ____

   ____

   ____

   C. Section 3: ____

   ____

   ____

   D. Section 4: ____

   ____

   ____

   E. Section 5: ____

   ____

   ____

Section Three: The Conclusion

8. What ideas do you have for concluding your speech?

____

____

____

9. How do you plan to "tell them what you told them"?

____

____

____

Name ______________________________

Gather the Facts

10. List the facts that you will be using for each of the sections. List at least three facts per section and include a source for each fact.

A. **Section 1: The Introduction**

1. Fact: ______________________________

____

Source: ______________________________

2. Fact: ______________________________

____

Source: ______________________________

3. Fact: ______________________________

____

Source: ______________________________

B. **Section 2: The Body**

1. Fact: ______________________________

____

Source: ______________________________

2. Fact: ______________________________

____

Source: ______________________________

3. Fact: ______________________________

____

Source: ______________________________

4. Fact: ______________________________

____

Source: ______________________________

5. Fact: ______________________________

____

Source: ______________________________

6. Fact: ______________________________

____

Source: ______________________________

C. **Section 3: The Conclusion**

1. Fact: ____

____

Source: ____

2. Fact: ____

____

Source: ____

3. Fact: ____

____

Source: ____

4. Fact: ____

____

Source: ____

Write Your Speech!

Use the information you have gathered to guide the development of your speech. Bring the facts for each section together. Use connections to your life or your audience's experiences to make the content more meaningful.

- The speech must be typed.
- The speech must be at least 2 1/2 pages (double-spaced).
- The topic may be anything related to agriculture.
- The speech should be approximately 5 minutes in length.
- Use proper spelling and punctuation.

Name ______________________________ Date ______________ Period __________

Activity 2C

Career Inventory

Objectives

- Discover your personal preferences for a career.
- Explore different career paths.

Materials

- activity worksheet
- internet access

Instructions

There are some guiding questions that can tell you a lot about your personal preference for a career. Complete the following chart to help you determine what you want in a career. For each category, circle the statement that best represents your preference.

Category A	Option A	Option B	Option C
Working environment	I want to be inside almost all of the time	I would like to have the freedom to be inside or outside	I want to spend most of my time outside
People or things	I want to spend most of my working time interacting with people	I want to work with people and things (products, inventory, etc.)	I want to work mainly with things (products, inventory, commodities, etc.)
Alone or with people	I want to spend most of my time working alone	I want to have some interaction with people during the day	I want to spend most of my workday working with other people
Geography	I want to be able to work close to where I grew up	I have a specific region in the country/world where I want to live and work	I am open to living and working anywhere
Salary	Making a lot of money is the most important factor to consider in choosing a career	I would like to make a lot of money, but I am willing to trade some money for the happiness of doing a job I love	Money is not that important to me, as long as I am able to do something I love
Education	I am willing to spend 4, 6, or even 8 years after high school graduation to get the schooling I need	I may be willing to spend 2 to 4 years after high school graduation to get the schooling I need	I only want to pursue education that will directly pertain to the skills needed for a particular career
Mind or body	I want to do work that mainly requires me to think about solutions	I want to be in a career where I can think about solutions, but spend time working with my hands	I want to spend most of my time at work being hands-on, accomplishing tasks, building things, or completing physical tasks

Goodheart-Willcox Publisher

Research Careers

Once you have completed the table, look through the career connections in your textbook or search online to find a job that meets your preferences. Write answers for each of the categories related to this specific career.

The career I chose is: ____________________

1. **Working environment**

   A. In what type(s) of environments is this work conducted? Will you spend most of your time indoors or outdoors? Is it a hot or cold environment? What other characteristics does the work environment have?

   ____

   ____

   ____

   B. How does this career match your selection in the preceding chart?

   ____

   ____

   ____

2. **People or things**

   A. What is the main focus of this career? Explain how the career focuses on either developing people or developing things.

   ____

   ____

   ____

   B. How does this career match your selection in the preceding chart?

   ____

   ____

   ____

3. **Alone or with people**

   A. How much time will you spend alone in this career?

   ____

   ____

   ____

   B. How does this career match your selection in the preceding chart?

   ____

   ____

   ____

Name ____

4. **Geography**

A. Where in the country/world will you be able to work with this type of career?

____

____

____

B. How does this career match your selection in the preceding chart?

____

____

____

5. **Salary**

A. What is the average starting salary for someone in this career?

____

____

____

B. How does this career match your selection in the preceding chart?

____

____

____

6. **Education**

A. How much education is required for this job after high school graduation?

____

____

____

B. How does this career match your selection in the preceding chart?

____

____

____

7. **Mind or body**

A. What percentage of your time will be spent thinking? What percentage of your time will be spent doing?

____

____

____

B. How does this career match your selection in the preceding chart?

____

____

____

Notes

Name ______________________ Date ____________ Period ________

Activity 2D

Leaders in History

Objectives

- Widen your knowledge of historical leaders.
- Identify leadership skills of successful people.
- Improve research, writing, and presentation skills.
- Identify personal strengths and weaknesses of your leadership skills.

Materials

- activity worksheet
- internet access
- note cards, paper
- pen

Instructions

You will have the opportunity to select someone from history who you consider a leader. What makes that person a leader? You will conduct a search of that person to identify their personal leadership traits, and look for insight into their life experiences which led them to become a leader. Review the list of leadership traits in your textbook to refresh your memory. After conducting the research, you will be responsible for developing a short presentation to share the information you gather with the rest of the class.

Leader you have selected: ______________________

1. List the top five leadership characteristics you feel this leader displayed. Share evidence of how your leader demonstrated each of the five characteristics.

   A. Leadership characteristic: ______________________

   Evidence: ______________________

   ____

   ____

   ____

   B. Leadership characteristic: ______________________

   Evidence: ______________________

   ____

   ____

   ____

C. Leadership characteristic: ______________________________

Evidence: ______________________________

____

____

____

D. Leadership characteristic: ______________________________

Evidence: ______________________________

____

____

____

E. Leadership characteristic: ______________________________

Evidence: ______________________________

____

____

____

2. Find information about your leader's life (background, childhood, experiences). What are the factors that you think might have led to the development of the characteristics you listed?

____

____

____

____

____

____

____

____

____

3. List at least three things this leader was able to accomplish because of his or her leadership characteristics.

____

____

____

____

____

____

____

____

Name ____

4. Can you think of anything this person could have strengthened to become a better leader?

5. Why did this person appeal to you as a historical leader? Share at least one similarity and one difference you have with this person related to your leadership skills.

6. What can this person teach us about being a good leader? What are the lessons we can take from this person and translate into situations we face today?

7. Use the information you collected about your leader to create your presentation. Presentations should share the information learned through this research from the class. You should also be prepared to answer questions from the class about your leader.

Notes

Name ______________________ Date ____________ Period ____________

CHAPTER 3 Historical Perspectives of Natural Resources and Conservation

Chapter Review

Carefully study the chapter and then answer the following questions.

1. _____ Which of the following forced Americans to rethink their agricultural practices?
   A. Land erosion
   B. Extinction of plant species
   C. Extinction of animal species
   D. All of the above.

2. _____ A period of social and political activism in the late 1800s is referred to as the _____.
   A. Conservation Era
   B. Industrial Era
   C. Progressive Era
   D. Revolutionary Era

3. _____ The need to protect many of our natural areas to ensure their longevity gave birth to the _____.
   A. Conservation Movement
   B. Industrial Movement
   C. Progressive Movement
   D. Revolutionary Movement

4. _____ The nation's first national park is _____.
   A. Mackinac National Park
   B. Great Plains National Park
   C. Yellowstone National Park
   D. Shelterbelt National Park

5. _____ Which of the following was established in 1916 by President Woodrow Wilson?
   A. Soil Conservation Service
   B. US National Park Service
   C. Civilian Conservation Corps
   D. Shelterbelt Corps

6. _____ Which of the following made farmers realize the importance of implementing extensive soil conservation practices?
   A. The Great Plains Shelterbelt
   B. The Conservation Movement
   C. The Progressive Movement
   D. The Dust Bowl

7. _____ The Soil Conservation Service (SCS) worked with farmers to _____.
   A. provide incentives to fallow land
   B. install soil conservation structures
   C. educate them about soil conservation practices
   D. All of the above.

8. _____ *True or False?* Around two million trees were planted in the Great Plains Shelterbelt to provide lumber for building.

9. _____ *True or False?* After World War II, the use of fertilizers and chemical pesticides allowed farmers to increase yields and reduce crop losses.

10. _____ *True or False?* The chemical pesticide DDT is an effective pesticide that had little effect on the environment.

11. _____ When a chemical pesticide bioaccumulates, it _____.
   A. is less effective
   B. builds up in organisms
   C. degrades with little impact on the environment
   D. All of the above.

12. _____ *True or False?* Natural disasters are caused solely through natural phenomena.

13. _____ Relatively small oil spills can cause great harm, depending on the _____.
   A. environmental sensitivity
   B. season and location
   C. type of oil
   D. All of the above.

14. _____ In total, the estimated amount of crude oil from the Exxon Valdez that was recovered from the ocean and shoreline is _____.
   A. 10%
   B. 20%
   C. 65%
   D. 98%

15. _____ The Amazon rain forest _____.
   A. is found primarily in Peru
   B. is home to 10% of all known species
   C. has low biodiversity
   D. All of the above.

Name ____

16. _____ Which of the following is an act of Congress that established the first national park in 1872?

A. Forest Reserve Act

B. Yellowstone National Park Act

C. Lacey Act

D. Sikes Act

17. _____ Which of the following allows the US President to set aside forest reserves from the land in public domain?

A. Forest Reserve Act

B. Yellowstone National Park Act

C. Dingell-Johnson Act

D. Sikes Act

18. _____ Which of the following was enacted to protect wildlife?

A. Forest Reserve Act

B. Yellowstone National Park Act

C. Lacey Act

D. Pitman-Robertson Act

19. _____ Which of the following was enacted to fund each state's management of endangered wildlife?

A. Dingell-Johnson Act

B. Endangered Species Act

C. Pitman-Robertson Act

D. Yellowstone National Park Act

20. _____ Which of the following authorizes the Secretary of the Interior to provide financial assistance for state fish restoration, projects, and management plans?

A. Magnuson-Stevens Act

B. Dingell-Johnson Act

C. Lacey Act

D. Sikes Act

21. _____ Which of the following ensures that the natural resources on military lands are protected and enhanced while allowing the land to meet the needs of the military?

A. Sikes Act

B. Dingell-Johnson Act

C. Lacey Act

D. Pitman-Robertson Act

22. _____ *True or False?* The first legislation to control air pollution is the Clean Forest Act.

23. _____ The Clean Water Act was passed to help _____.
A. provide funds for municipal waste treatment
B. eliminate water pollution
C. preserve natural wetlands
D. All of the above.

24. _____ Which of the following was enacted to regulate fishing in US territorial waters?
A. Dingell-Johnson Act
B. Magnuson-Stevens Act
C. Pitman-Robertson Act
D. Lacey Act

25. _____ Ducks Unlimited, the National Wild Turkey Federation, the Sierra Club, and the Audubon Society are _____.
A. government organizations
B. state organizations
C. advocacy organizations
D. All of the above.

Vocabulary Review

Match each conservationist with the appropriate description.

A. Norman Borlaug
B. Archie Carr
C. Rachel Carson
D. Jacques-Yves Cousteau
E. Dian Fossey
F. Jane Goodall
G. Aldo Leopold
H. John Muir
I. Margaret Murie
J. Gaylord Nelson
K. Gifford Pinchot
L. Theodore Roosevelt
M. George Vanderbilt
N. E.O. Wilson
O. Charles Young

1. _____ A wealthy conservationist who repaired the worn-out, heavily forested land around his estate and established the Biltmore Forest School.
2. _____ A biologist and tropical field ecologist helped save the sea turtle from extinction.
3. _____ A primatologist who studied chimpanzees for more than 40 years.
4. _____ A marine biologist credited with helping ban DDT use in the United States.
5. _____ A plant biologist who developed high-yielding, disease-resistant wheat varieties.
6. _____ A primatologist who studied and campaigned for protection of mountain gorillas.
7. _____ A researcher whose documentary films and television series helped educate viewers to ocean life and conservation.
8. _____ One of the first leaders to realize our natural resources were not inexhaustible and who established the US Forest Service.

Name ______________________________

9. _____ A senator and conservation advocate who founded Earth Day.

10. _____ After serving 12 years in the cavalry, he was appointed superintendent of Sequoia and General Grant National Parks.

11. _____ An American biologist who is recognized as the world's leading authority on ants.

12. _____ Studied elk and other wildlife of the Yellowstone-Grand Teton areas and worked with fellow conservationists to advocate conservation of our natural areas.

13. _____ An ecologist and professor who helped establish the first wilderness area in the US Forest Service and wrote a book that advocates land ethic.

14. _____ Conservationist credited with fostering the idea of conservation ethic and the first chief of the US Forest Service.

15. _____ One of the petitioners of the original park legislation that created Yosemite National Park and who is referred to as the "Father of National Parks."

Match each government organization with the appropriate description.

A. BLM
B. DOI
C. NMFS
D. NOAA
E. NRCS
F. USFS
G. USFWS

16. _____ Oversees nearly 200 million acres of national forests and grasslands.

17. _____ Protects endangered species, manages migratory bird habitats, protects fish sanctuaries, and restores wetlands for wildlife.

18. _____ Responsible for the stewardship of the nation's marine resources within the US Exclusive Economic Zone.

19. _____ Created in 1946 and manages almost 250 million acres of public land.

20. _____ Provides technical and financial assistance to establish conservation programs on private and public lands.

21. _____ Provides various public services, including daily weather forecasts, severe storm warnings, coastal restoration, support to marine commerce, and conducts research to improve stewardship of the environment.

22. _____ Responsible for management and conservation of federal lands, as well as programs related to Native Americans.

Thinking Critically

1. Describe what occurred in 2010 on the British Petroleum Deepwater Horizon oil rig. Explain the impact this had on the environment.

2. Choose one or more of the true/false questions that you marked as false and explain why you marked it false.

3. Identify a piece of literature that discusses an environmental issue and explain how it can serve as a vehicle for social change.

Name ______________________ Date ____________ Period ________

Activity 3A

People Who Make a Difference

Objectives

- Expand your knowledge of environmentalists and conservationists.
- Improve your research, reading, and writing skills.
- Write a concise summary of a famous environmentalist or conservationist.

Materials

- library access
- computer and printer access

Environmentalists

Many people have championed the conservation and preservation of our environment. Can you picture what many of our preserved areas may have become if these people had not fought for legislation to establish our national protected areas? Choose one person from the following list and research his or her life and work. Write a one-page summary on the life of the environmentalist/conservationist that you selected. All research is to be done using a minimum of three sources of printed materials. You may not *cite information from online sources. The second page must include all works cited in MLA format. Your summary must include the following information:*

- ***Who*** the person is and some background and/or history of the person (10 points)
- ***What*** the person is noted for doing and why they are important to their specific cause(s) (15 points)
- ***When*** the person lived and the time period in which they did their work (10 points)
- ***Where*** the person conducted work and where their research went afterwards (10 points)
- ***Why*** the research and/or work was needed and its impact (10 points)
- ***Why*** he or she is considered to be an environmentalist or conservationist (10 points)
- ***How*** the person made an impact or is making an impact on the world (15 points)
- One photo of the person (the image may be obtained from an online source) (5 points)
- Works cited in MLA format (3 source minimum) (10 points)
- Spelling and grammar. You must use complete sentences with correct spelling and proper punctuation (5 points)

Edward Abbey

David Attenborough

Ed Begley Jr.

Norman Borlaug

Erin Brockovich

Harvey Broome

Pierce Brosnan

David Brower

Lester R. Brown

Robert Bullard

Archie Carr

Rachel Carson

Jimmy Carter

George Washington Carver

Jeff Corwin

Jacques-Yves Cousteau

Rosalie Edge

Diann Fossey

Maneka Gandhi

Jane Goodall

Al Gore

Daryl Hannah

Julia Hill

Winona LaDuke

Aldo Leopold

James Lovelock

Amory Lovins

Wangari Maathai

Jamie Margolin

Robert Marshall

Chico Mendes

John Muir

Margaret Murie

Olaus Murie

Gaylord Nelson

Sigurd F. Olson

Gifford Pinchot

Theodore Roosevelt

Vandana Shiva

Hilda Lucia Solis

David Suzuki

Henry David Thoreau

Greta Thunberg

George Vanderbilt

Malaika Vaz

Penny Whetton

E.O. Wilson

Robert Sterling Yard

Charles Young

Howard Zahniser

Name ______________________________ Date ______________ Period __________

Activity 3B

Conservation Time Line

Objectives

- Identify key changes in conservation practices.
- Improve research and writing skills.
- Create a comprehensive classroom display.

Materials

- activity worksheet
- internet access
- printer
- poster board (one per group)

Instructions

Your instructor will divide your class into several groups. Each group will create a time line with images that show how conservation has changed in the last 150 years. Find at least one image showing conservation practices or major conservation events for each 20-year section. Print the photos and create a time line on the poster board beginning with 1870 and ending today. Label the images. Use this worksheet to collect information to accompany your photo time line. Type your information in the appropriate type size for viewing at a short distance. Cut the information and attach it to the poster board.

1. **1870–1890**

   A. Image identification/description: ______________________________

   ____

   ____

   B. How does the subject qualify as conservation?

   ____

   ____

   ____

2. **1890–1910**

   A. Image identification/description: ______________________________

   ____

   ____

   B. How does the subject qualify as conservation?

   ____

   ____

   ____

3. **1910–1930**

A. Image identification/description: ______________________

____

____

B. How does the subject qualify as conservation?

____

____

____

4. **1930–1950**

A. Image identification/description: ______________________

____

____

B. How does the subject qualify as conservation?

____

____

____

5. **1950–1970**

A. Image identification/description: ______________________

____

____

B. How does the subject qualify as conservation?

____

____

____

6. **1970–1990**

A. Image identification/description: ______________________

____

____

B. How does the subject qualify as conservation?

____

____

____

Name ____________________

7. **1990–2010**

A. Image identification/description: ____________________

B. How does the subject qualify as conservation?

8. **2010–2020**

A. Image identification/description: ____________________

B. How does the subject qualify as conservation?

9. **Present**

A. Image identification/description: ____________________

B. How does the subject qualify as conservation?

10. Which time period do you feel had the greatest impact on conservation? Why?

Notes

Name ______________________________ Date ______________ Period __________

CHAPTER 4 Ecology and Earth

Chapter Review

Carefully study the chapter and then answer the following questions.

1. _____ *True or False?* A biosphere is a biological community of living beings within an environment or similar climatic area.

2. _____ An ecosystem contains _____.
   A. soil and minerals
   B. plants, animals, and microorganisms
   C. air and water
   D. All of the above.

3. _____ The main factors that determine or affect the climate of an area are temperature, air pressure, humidity, winds, daylight, and _____.
   A. precipitation
   B. minerals
   C. microorganisms
   D. All of the above.

4. _____ *True or False?* The surface configuration or features of a land area have little effect on the types of animals and plants that inhabit an area.

5. _____ The primary mineral content of soil is contributed through the soil's _____.
   A. microorganisms
   B. water supply
   C. parent material
   D. All of the above.

6. _____ *True or False?* The water cycle is the pathway that water follows at, above, and below Earth's surface.

7. _____ The process through which atmospheric nitrogen is transformed into ammonium is _____.
   A. nitrification
   B. ammonification
   C. denitrification
   D. fixation

8. _____ The process through which bacteria convert organic nitrogen into ammonium is _____.
   A. nitrification
   B. ammonification
   C. denitrification
   D. fixation

9. _____ The process of converting ammonium into nitrates using nitrifying bacteria is _____.
   A. nitrification
   B. ammonification
   C. denitrification
   D. fixation

10. _____ The conversion of nitrates into inert atmospheric nitrogen in _____.
   A. nitrification
   B. ammonification
   C. denitrification
   D. fixation

11. _____ *True or False?* Carbon is a basic building block for plant and animal life.

12. _____ *True or False?* The energy cycle is the movement of energy through a living being.

13. _____ A large grouping of communities of distinctive flora and fauna living in a particular region is a(n)_____.
   A. eco-zone
   B. ecoregion
   C. biome
   D. All of the above.

14. _____ An area of similar ecosystems in the type, quality, and quantity of natural resources is a(n) _____.
   A. eco-zone
   B. ecoregion
   C. biome
   D. All of the above.

15. _____ Biomes that include mangrove, coral, and deepwater ecosystems are _____.
   A. lake biomes
   B. tundra biomes
   C. aquatic biomes
   D. rain biomes

16. _____ Biomes that are divided into hot and dry, semiarid, coastal, and cold are _____.
   A. desert biomes
   B. tundra biomes
   C. sand biomes
   D. dry biomes

17. _____ Rain forests, temperate forests, and taiga forests are _____.
   A. tundra biomes
   B. forest biomes
   C. tree biomes
   D. tropical biomes

18. _____ The prairies in North America, the pampas in South America, the steppes in Asia, and the veld in Southern Africa are _____.
   A. tundra biomes
   B. prairie biomes
   C. grassland biomes
   D. plain biomes

19. _____ Arctic, alpine, and Antarctic biomes are _____.
   A. anthropogenic biomes
   B. grassland biomes
   C. winter biomes
   D. tundra biomes

20. _____ Terrestrial ecoregions that have been altered by agriculture, urbanization, and other human activities are _____.
   A. plain ecoregions
   B. anthropogenic ecoregions
   C. forest ecoregions
   D. agricultural ecoregions

21. _____ Which of the following statements is true about urbanization?
   A. Precipitation runs off quickly instead of slowly infiltrating the soil
   B. Hard surfaces absorb heat during the day and release it at night
   C. Air conditioners release large volumes of heat
   D. All of the above.

22. _____ Which of the following are included in freshwater ecosystems?
   A. Inland wetlands
   B. Springs, rivers, and streams
   C. Ponds and lakes
   D. All of the above.

23. _____ *True or False?* Lentic bodies of water include rivers and streams.

24. _____ *True or False?* Most lakes support a variety of aquatic plant materials because sunlight easily reaches the bottom of the lake.

For questions 25–33, identify the zones and types of plants found in ponds and lakes indicated in the following illustration.

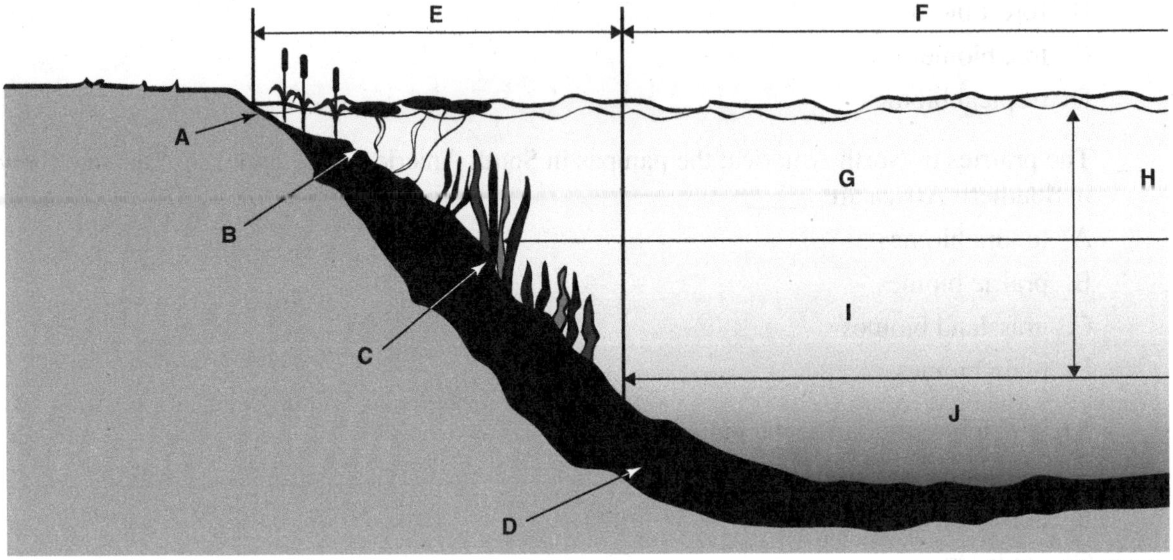

Goodheart-Willcox Publisher

25. _____ benthic zone

26. _____ emerged/erect plants

27. _____ epilimnion zone

28. _____ euphotic zone

29. _____ floating plants

30. _____ limnetic zone

31. _____ littoral zone

32. _____ oxygenating plants

33. _____ profundal zone

Name ______________________________

34. _____ *True or False?* Moving bodies of water are lotic.

35. _____ Streams and rivers begin at a location referred to as the _____.

A. headwaters

B. tributary

C. delta

D. main channel

36. _____ The most abundant and diverse species in rivers are found in the _____.

A. tributary zone

B. thermo zone

C. transition zone

D. All of the above.

37. _____ Which of the following is *not true* about the floodplain zone of a river?

A. Its water is clearer and colder than the water in other zones

B. It has more sediment than other zones

C. It is warmer and has less oxygen than other zones

D. It contains more nutrients than other zones

38. _____ The areas where freshwater rivers and streams merge with the ocean and mix with saltwater are _____.

A. headwaters

B. estuaries

C. lagoons

D. deltas

39. _____ Which of the following wetlands are found along the Alaskan, Atlantic, Pacific, and Gulf coasts?

A. Compositional wetlands

B. Nontidal wetlands

C. Tidal wetlands

D. Benthic wetlands

40. _____ Which of the following wetlands are found along rivers and streams or in isolated areas surrounded by dry land?

A. Compositional wetlands

B. Nontidal wetlands

C. Tidal wetlands

D. Benthic wetlands

41. _____ *True or False?* Brackish water is a combination of freshwater and saltwater.

42. _____ *True or False?* Saltwater accounts for about 3% of the water on Earth.

Vocabulary Review

Match each of the components of the carbon cycle with its appropriate description.

A. carbon source
B. carbon pool
C. carbon sink

1. _____ It releases more carbon than it stores.
2. _____ An area in which carbon accumulates and is also released.
3. _____ An area that stores carbon for long periods.

Match each type of ecoregion with its appropriate description.

A. eastern temperate forest ecoregion
B. Great Plains ecoregion
C. marine west coast forest ecoregion
D. Mediterranean California ecoregion
E. North American desert ecoregion
F. northern forests ecoregion
G. northwestern forested mountains ecoregion
H. temperate Sierras ecoregion
I. tropical wet forest ecoregion

4. _____ An ecoregion that is practically devoid of water much of the year.
5. _____ The climate of this ecoregion is warm throughout most of the year with abundant rainfall in the fall, winter, and spring.
6. _____ An ecoregion that is primarily chains of mountains covered with dense coniferous forests and sagebrush.
7. _____ Precipitation falls more densely in the ecoregion than anywhere else in the United States.
8. _____ Tall trees and woody shrubs dominate this ecoregion, which is the largest of the American ecoregions.
9. _____ The defining plant life of this ecoregion is the prairie tallgrass.
10. _____ This ecoregion has the most diverse conifer forests on Earth.
11. _____ An ecoregion with tall trees, abundant rainfall, and an abundance of wildlife.
12. _____ This ecoregion, which extends primarily across Canada, has long, cold winters with short, warm summers.

Name ______________________________

Match each of the zones with its appropriate description.

A. benthic zone
B. limnetic zone
C. littoral zone
D. profundal zone

13. _____ The open water near the surface of a pond or lake.

14. _____ The deeper, cooler, darker zone of water in which few fish species live.

15. _____ The bottom sediment which has a surface layer abundant with organisms.

16. _____ The shallow, warm, and well-lit zone that is close to the shore.

Thinking Critically

1. What is the driving force of the water cycle?

____
____
____
____
____
____

2. What is the path of phosphorus?

____
____
____
____
____
____
____

3. Choose an aquatic animal and explain how changes in the water cycle affect this animal.

____
____
____
____
____
____
____

Notes

Name ______________________ Date ____________ Period ________

Activity 4A

Biogeochemical Cycles

Objectives

- Describe the series of steps in each biogeochemical cycle.
- Explain why each cycle is important to our environment and natural resources.

Materials

- activity worksheet
- internet access

Instructions

Demonstrate your understanding of the biogeochemical cycles discussed in the textbook by briefly explaining what occurs in each cycle and why it is important to our environment and natural resources.

The Water Cycle

1. **The Water or Hydrologic Cycle**

   A. What occurs during the water or hydrologic cycle?

   ____

   ____

   ____

   ____

   ____

   B. Why is the water cycle important to our environment and natural resources?

   ____

   ____

   ____

   ____

   ____

2. **The Nitrogen Cycle**

   A. What occurs during the nitrogen cycle?

   ____

   ____

   ____

   ____

   ____

B. Why is the nitrogen cycle important to our environment and natural resources?

3. **The Carbon Cycle**

A. What occurs during the carbon cycle?

B. Why is the carbon cycle important to our environment and natural resources?

4. **The Energy Cycle**

A. What occurs during the energy cycle?

B. Why is the energy cycle important to our environment and natural resources?

5. Which cycle do you think is most important? Explain your answer.

Name ______________________________ Date ______________ Period __________

CHAPTER 5 Sustainability in the Environment

Chapter Review

Carefully study the chapter and then answer the following questions.

1. _____ Meeting the needs of the present without compromising the ability of future generations to meet their own needs is referred to as _____.
   A. rationality
   B. suitability
   C. sustainability
   D. polarity

2. _____ The preservation of humankind and quality of life is _____.
   A. economic sustainability
   B. social sustainability
   C. ecological sustainability
   D. anthropogenic sustainability

3. _____ Which of the following is the practice of maintaining biological systems so that they remain diverse and productive?
   A. Economic sustainability
   B. Social sustainability
   C. Environmental sustainability
   D. Anthropogenic sustainability

4. _____ Which of the following is sustainability in which a society maintains sufficient resources to support a specific level of production, indefinitely?
   A. Economic sustainability
   B. Social sustainability
   C. Ecological sustainability
   D. Anthropogenic sustainability

5. _____ The production of a natural resource using procedures which ensure replacement of the portion harvested by regrowth before another harvest occurs is _____.
   A. harvest yield
   B. sustained yield
   C. unsustainable yield
   D. replacement yield

6. _____ The approach we take and the processes we use to achieve sustainability are referred to as sustainable _____.

A. construction
B. growth
C. development
D. All of the above.

7. _____ The production of food, fiber, and forest products using farming techniques that protect the environment, public health, and social communities is sustainable _____.

A. development
B. forestry
C. viability
D. agriculture

8. _____ Which of the following are sustainable agriculture practices?

A. Minimizing water use
B. Improved seed technology and genetics
C. Drones monitoring crops
D. All of the above.

9. _____ An energy whose consumption is less than the supply replenishment and that has negligible impact on the environment is _____.

A. sustainable energy
B. fossil energy
C. kinetic energy
D. reusable energy

10. _____ Which of the following lightbulbs is the most energy-efficient?

A. Compact fluorescent
B. LEDs
C. Halogen
D. Incandescent

11. _____ *True or False?* Using a programmable thermostat to raise and lower the temperature when no one is home will reduce home energy consumption.

12. _____ The design and construction of structures using sustainable technology and materials is _____.

A. xeriscaping
B. green building
C. LED construction
D. All of the above.

13. _____ *True or False?* Water-smart buildings use technology such as low-flow toilets and gray water from washing machines and showers to flush toilets.

Name ______________________________

14. _____ Which of the following statements about xeriscaping is true?
   A. It uses plants native to the area.
   B. It reduces the amount of lawn that needs to be irrigated.
   C. It uses less water than traditional landscaping.
   D. All of the above.

15. _____ *True or False?* A green roof gets its name from the color of the environmentally safe green paint used to paint it.

Vocabulary Review

Match the following terms with the appropriate definition or description.

A. compost
B. gray water
C. hazardous waste
D. landfill
E. liquid waste
F. recycled
G. solid waste
H. upcycled
I. waste disposal
J. wastewater
K. zero waste

1. _____ When a product is repurposed and increases in value.
2. _____ An engineered construction project designed to minimize the amount of materials that are thrown in it.
3. _____ Contains the most mass of all waste products and uses the most disposal space.
4. _____ Water from bathing, cooking, and general household use.
5. _____ When a product is modified in some way and turned into a usable product.
6. _____ Decaying organic matter that can be used as a soil amendment.
7. _____ Waste such as contaminated water, used oil, and residual chemicals.
8. _____ The storage, removal, or destruction of unwanted domestic, agricultural, and industrial products.
9. _____ A philosophy where people strive to either use or recycle all resources and send nothing to the landfill or incinerator.
10. _____ Toxic, ignitable, corrosive, or reactive waste that poses a substantial threat.
11. _____ Water that has been used and is no longer clean.

Thinking Critically

1. Explain why the early colonists were not concerned with the availability of natural resources.

2. What is the purpose of the National Environmental Policy Act?

3. Why is purchasing locally grown food good for the environment?

Name ____

4. Make a list of your environmental values, such as what you think is right or wrong environmentally. How can you put your values into action?

5. Do our values affect science? How can science affect our values?

6. Is the human race superior to other living things? How has this belief influenced the designs of human systems, such as power generation and agriculture?

Notes

Name ____________________ Date ____________ Period ________

Activity 5A

Energy Usage Lab

Objective

- Understand the need for renewable energy sources.

Materials

- activity worksheet
- two different colors of beads, beans, or candy for each group (80 of one color, 20 of the other)
- two bowls for each group

Instructions

According to the US Department of Energy, 80% of our energy consumption is from nonrenewable resources, and 20% is from renewable resources. Do those numbers surprise you? Can we change those numbers? This lab will help you understand the need for renewable energy sources. Follow each of the steps with your partner(s) and answer the questions.

Section 1: Constant Usage

1. Place 80 items (beads/beans/candy) of one color and 20 of another color in one of your bowls. The 80 items represent the 80% of US energy usage that comes from nonrenewable sources. The 20 items represent the 20% of US energy usage that comes from renewable energy sources.
2. Have one partner close his or her eyes and randomly draw 10 items from the first bowl and place them in the empty bowl. Write down (in the row for round one on the chart) the number of renewable and nonrenewable items that your partner drew.
3. Sort through the items and return all the renewable items to the original bowl. Leave the nonrenewable items picked in the first round in the second bowl.

4. Repeat the process for an additional 10 rounds. Record the numbers of each color drawn and replace the renewable items after each drawing. The number of nonrenewable items in the original bowl will decrease with each round. Record the information in the following table.

Create an Energy Usage Ratio			
Round	**# To Draw**	**# Renewable**	**# Nonrenewble**
1	10		
2	10		
3	10		
4	10		
5	10		
6	10		
7	10		
8	10		
9	10		
10	10		
11	10		

Goodheart-Willcox Publisher

5. How many of each type of item are left in the original bowl after the final round?

   A. Renewable items: ____________________

   B. Nonrenewable items: ____________________

6. How many rounds do you think it would take before all of the nonrenewable items were gone?

   Rounds: ____________________

7. How does this section of the lab relate to the use of renewable versus nonrenewable resources in the United States?

____

____

____

____

____

____

____

____

Name ____________________

Section 2: Increasing Use

1. Place 75 items (beads/beans/candy) of one color and 20 of another color in one of your bowls. The 75 items represent the 75% of US energy usage that comes from nonrenewable sources. The 20 items represent the 25% of US energy usage that comes from renewable energy sources.
2. Have one partner close his or her eyes and randomly draw 10 items from the first bowl and place them in the empty bowl. Write down (in the row for round one on the chart) the number of renewable and nonrenewable items that your partner drew.
3. Sort through the items and return all the renewable items to the original bowl. Leave the nonrenewable items picked in the first round in the second bowl.
4. Repeat the process for an additional 10 rounds, but this time, draw an *additional item* each round (i.e., draw 11 items in round 2, 12 in round 3, etc.). Record the numbers of each color drawn and replace the renewable items after each drawing. The number of nonrenewable items in the original bowl will decrease with each round. Record the information on the following table.

Create an Energy Usage Ratio

Round	# To Draw	# Renewable	# Nonrenewble
1	10		
2	11		
3	12		
4	13		
5	14		
6	15		
7	16		
8	17		
9	18		
10	19		
11	20		

Goodheart-Willcox Publisher

5. How many of each type of item are left in the original bowl after the final round?

   A. Renewable items: ____________________

   B. Nonrenewable items: ____________________

6. How many rounds do you think it would take before all of the nonrenewable items were gone?

   Rounds: ____________________

7. How does this section of the lab relate to the use of renewable versus nonrenewable resources in the United States?

8. What does the difference between the two rounds indicate about the need for both increasing renewable energy sources and conserving energy sources to prevent increased usage?

Name ______________________ Date ______________ Period __________

Activity 5B

Biofuel Lab

Objective

- Understand the basic process for making a biofuel.

Materials

- activity worksheet
- (2) 250 ml graduated cylinders
- (2) 250 ml beakers
- balance
- canola oil
- chemical-resistant apron
- glass stirring rod
- goggles
- hot plate
- methanol
- potassium hydroxide (KOH)
- spatula
- thermal gloves
- thermometer
- weighing paper
- wide-mouthed jar with lid

SAFETY NOTE

The lab activity should be conducted in a science classroom with splash guards, an eye wash station, and basic lab equipment.

1. Chemicals are used during this lab. Wear chemical splash goggles, a chemical-resistant apron, and chemical-resistant rubber gloves during the entire procedure. Avoid contact between skin and liquids.
2. Label all containers with their contents.
3. Work cautiously with hot oil to prevent burns.
4. Follow instructions for proper storage and/or disposal of chemicals.

Day One

1. Put on your personal protective equipment (goggles, apron, thermal gloves).
2. Mix the methanol and potassium hydroxide (KOH).
   - Measure 20 ml of methanol in a graduated cylinder. Pour the methanol into a 250 ml beaker.
   - Measure 0.7 grams of KOH and add to the beaker of methanol.
   - Stir slowly until the KOH is completely dissolved (8–10 minutes of stirring may be necessary).

CAUTION

Do not splash this mixture as it is very harmful to skin.

3. Heat the canola oil.
   - Measure 100 ml of canola oil in a graduated cylinder. Pour the canola oil into a 250 ml beaker. Note: This is not the same beaker containing the methanol and KOH mixture.
   - Place the beaker with the canola oil on a hot plate. Place a thermometer in the canola oil and heat until the temperature reaches 120°C. Note: Keep the beaker with the methanol and KOH mixture away from the hot plate.
   - When the oil temperature reaches 120°C, remove the beaker and turn off the hot plate.
   - Carefully pour the heated oil into the wide-mouthed jar.
4. Add the methanol and potassium hydroxide mixture to the heated oil.
   - Slowly add the methanol and potassium hydroxide mixture to the heated oil in the jar. Put the lid on the jar and shake the mixture for 15 minutes. Caution: The jar may be warm or hot to the touch.
   - After you have shaken the mixture for 15 minutes, allow it to sit for 60 seconds.
   - Loosen the lid slowly to vent the jar and then retighten the lid.
   - This mixture is a combination of biofuel and glycerin.
5. Let the mixture sit overnight.
6. Clean your workstation.

Day Two

1. Put on your personal protective equipment (goggles, apron, thermal gloves).
2. Examine the separated biofuel and glycerin solution.
   - The darker liquid at the bottom is glycerin.
   - The clear liquid at the top is biofuel.
3. Decant the biofuel into a graduated cylinder. *Decanting* is the process of pouring a liquid from a sediment. In this lab, the liquid is the biofuel and the sediment is the glycerin.
   - Place the glass stirring rod against the point on the mouth of the jar where you will be pouring. (The stirring rod will help keep the glycerin from being poured in the jar.)
   - Carefully pour the biofuel from the jar into the graduated cylinder. The liquid will run down the stirring rod into the cylinder. Take care to not pour the glycerin into the cylinder.
4. Ask your teacher for instructions on the handling and storage of your biofuel.
5. Clean your workstation.

Name ______________________________ Date ______________ Period __________

Activity 5C

Carbon Footprints

Objectives

- Use the EPA online carbon footprint calculator.
- Calculate your personal carbon footprint.

Materials

- activity worksheet
- internet access
- printer and paper

Instructions

Follow the steps listed below and examine your carbon footprint.

1. Search for the EPA's carbon footprint calculator. Enter your information and complete the assessment of your carbon footprint. Once you receive your report, print it and answer the following questions.
2. What is your current usage?

   ____

3. What is your new total usage after planned activities?

   ____

4. What is the US average for your size household in your zip code?

   ____

5. What are three of the planned changes you agreed to in order to help reduce your carbon footprint?

   ____

   ____

   ____

6. What is the difference between your current carbon usage total and the total after your planned changes?

   ____

   ____

7. According to the data from your report, to what are your actions equal in regard to the following areas?

   A. Saving ______________ gallons of gas.

   B. Planting ______________ trees.

   C. Recycling ______________ tons of waste.

Notes

Name ____________________ Date ____________ Period ________

Activity 5D

The Organic Debate

Objectives

- Determine the economic differences of organic produce and conventional produce.
- Determine the pros and cons of organic versus conventional produce.
- Improve research, writing, and debate skills.

Materials Needed

- internet access
- activity worksheet

Instructions

There is a great deal of discussion about the differences between organic and conventionally-produced fruits and vegetables. In this activity, you will first examine the economic differences for consumers. In section two, you will research the benefits of each type of produce and prepare an argument in favor of one or the other.

Section One: Economic Differences

Visit a local supermarket or search for online grocery store prices. Write the organic and non-organic prices for the products listed in the chart. Calculate the dollar difference and the percentage difference.

Product	Organic Price per Pound	Nonorganic Price per Pound	$ Difference	% Difference
Carrots				
Apples				
Bananas				
Peppers				
Pears				
Strawberries				
Spinach				

Section Two: Benefits of Organic Produce

Define organic produce and search online for factual data on the benefits of organic produce. Write what you feel are five of the best supporting facts for raising, purchasing, and consuming organic produce. Include the source(s) for your data.

1. What is organic produce?

____

____

____

____

____

Five Best Supporting Facts

1. Supporting fact:

____

____

____

Source: ______________________________________

____

2. Supporting fact:

____

____

____

Source: ______________________________________

____

3. Supporting fact:

____

____

____

Source: ______________________________________

____

4. Supporting fact:

____

____

____

Source: ______________________________________

____

Name ______________________

5. Supporting fact:

Source:

6. What are the cons for raising, purchasing, and consuming organic produce?

Source:

Section Three: Benefits of Conventional Produce

Search online for factual data on the benefits of conventional (nonorganic) produce. Write what you feel are five of the best supporting facts. Include the source(s) for your data.

1. Supporting fact:

Source:

2. Supporting fact:

Source:

3. Supporting fact:

Source:

4. Supporting fact:

Source:

5. Supporting fact:

Source:

6. What are the cons for raising, purchasing, and consuming conventional produce?

Source:

Section Four: Form Your Argument

1. Write a two- to three-sentence argument for purchasing and consuming only organic produce.

2. Write a two- to three-sentence argument for purchasing and consuming only conventionally grown produce.

3. Use these arguments to prepare for a class debate on the topic of organic and conventionally produced foods.

Name ______________________ Date ______________ Period ________

CHAPTER 6 Habitat Destruction

Chapter Review

Carefully study the chapter and then answer the following questions.

1. _____ The natural environment a being needs to live is its _____.
   A. territory
   B. crib
   C. range
   D. habitat

2. _____ When a habitat is changed in such a way that it can no longer meet the needs of its inhabitants, it has experienced habitat _____.
   A. installation
   B. destruction
   C. reclamation
   D. annihilation

3. _____ *True or False?* A habitat's health can be evaluated by determining which organisms are thriving in the location.

4. _____ *True or False?* The least affected species are identified and studied to determine a habitat's health.

5. _____ *True or False?* Most natural habitat disturbances are caused by human actions.

6. _____ *True or False?* Habitat disturbances caused by wildlife is detrimental to the health of the ecosystem.

7. _____ Which of the following actions are taken to help remove or control invasive species?
   A. Removing common food sources
   B. Physically removing or harvesting the invasive species
   C. Mapping where the invasive species has been seen
   D. All of the above.

8. _____ The division of a large, continuous habitat into smaller, disconnected pieces is referred to as habitat _____.
   A. impaction
   B. degradation
   C. fragmentation
   D. installation

9. _____ The number of living organisms that a habitat can support without degradation is the land's _____.

A. carrying capacity
B. biodiversity limitation
C. habitat capacity
D. land limitation

10. _____ *True or False?* A fragmented habitat has more carrying capacity than a habitat that is intact.

11. _____ When a habitat is altered and made less suitable for the native wildlife, it is experiencing habitat _____.

A. impaction
B. degradation
C. fragmentation
D. installation

Vocabulary Review

Match the terrestrial biomes to the appropriate definition or description.

A. deciduous forest biome
B. desert biome
C. Mediterranean vegetation biome
D. montane biome
E. polar biome
F. savanna biome
G. steppe biome
H. taiga biome
I. tropical rain forest biome
J. tundra biome

1. _____ Includes the forests, grasslands, shrublands, and waterways situated in higher altitudes; found in mountain ranges around the world.

2. _____ A grassland with a very dry season and a very wet season.

3. _____ A vast region of cold, treeless lands with a permanently frozen subsoil.

4. _____ An area in which the annual precipitation is less than 9.8″ (250 mm) and the mean temperature during the warmest month of the year is less than 50°F (10°C).

5. _____ Consists of coniferous forests and is the largest biome (aside from the oceans) in the world.

6. _____ An area that is halfway between a forest and a desert; typically, treeless prairie or grasslands that experience seasonal changes.

7. _____ Area with annual rainfalls of at least 100″ (254 cm) and temperatures averaging around 80°F (27°C).

8. _____ The driest locations on the planet, receiving less than 10″ (25 cm) of rainfall each year.

9. _____ Areas characterized by hot, dry summers and mild wet winters.

10. _____ Areas that experience four seasons in which the majority of the trees and shrubs shed their leaves during the cold season.

Name ______________________________

Thinking Critically

1. What should we be doing now to help stop habitat destruction?

2. What are governments doing to protect biodiversity?

3. Can we create a balance of conservation and development that will prevent habitat loss? Explain your answer.

Notes

Name ______________________ Date ______________ Period __________

Activity 6A

Xeriscape Design Plan

Objectives

- Determine best plant types for a local xeriscape design.
- Design a xeriscape for a local property.
- Improve measuring and drawing skills.

Materials

- colored pencils or markers
- drafting pencils
- eraser
- graph paper
- internet access
- measuring wheel, long tape, or phone app for measuring
- poster board or drafting paper
- rulers and other drawing tools (compass, stencils, protractor)

Xeriscaping

Xeriscaping is landscaping that reduces or eliminates the need for irrigation. Many people use plants that are native to the area and those that require little water. Most xeriscapes feature various hardscapes, such as stone walkways, large accent boulders, and stone mulch. In this activity, you will design a xeriscape for a specific property.

Instructions

Research plant material recommendations for your area. You should also review plant-specific planting distances, shade/light preferences, and irrigation needs. A visit to a local nursery or garden center may also be of assistance.

Step One: Measuring the Property

1. Choose a property. You may redesign your home's landscape, choose a site on school grounds, or select another property.
2. Quickly sketch an aerial view of the property. It may prove useful to quickly photograph the property from different viewpoints for later reference.
3. Measure the property and note measurements on your sketch. If you are unable to obtain actual measurements, you may use estimated measurements.

Step Two: Plants and Hardscapes

Your landscape design must include the following:

- An aerial view of the property lines and existing buildings (available online)
- Hardscape features such as driveways, walkways, decks, and playground areas (existing features may be used if they are to be incorporated into the design)
- At least one hardscape that is somewhat permeable (allows water to drain through it)
- Flowering plants, 1–3 trees (existing trees may be used), a live ground cover (other than turfgrass), two different kinds of shrubs, and any other plants that will suit your xeriscape
- Garden ornaments, solar lights, and/or structures for climbing plants (optional)

Step Three: Drawing Your Design

If you are unfamiliar with landscape drawings, search online for tutorials and examples of icons, title blocks, and other features used in landscape designs. If you have access to or are familiar with a drafting program, you may use it to draw your xeriscape.

1. If you are not using prepared drafting paper, draw an outline on the paper and draw a title block with your name, date, and project title. Include the scale if you are drawing your plan to scale.
2. Using your original sketch as a guide, accurately draw the property on the poster board or drafting paper. Draw the property lines, buildings, and existing hardscape features first.
3. Draw any new features you are using, such as patios or walkways.
4. Draw planting area outlines as needed and then draw plant materials in their desired locations. You may search online for examples of common landscape design icons. You may also include a key for your drawing that includes all materials used.
5. Neatly label the components on your drawing. You may use abbreviations to avoid crowding if you have created a key.
6. Use colored pencils or markers to complete your drawing.
7. If time allows, compile a list of materials that includes prices.

Name ______________________ Date ____________ Period ________

Activity 6B

Carrying Capacity

Objectives

- Determine how limiting factors play a role in animal survival and behavior in an ecosystem.
- Understand how limited resources affects the carrying capacity of an ecosystem.

Materials

- papers labeled "water" (1 for each student in the class)
- papers labeled "food" (1 for each student in the class)
- large area for students to move around in (gym, soccer field, etc.)

Instructions

Your teacher will place food and water resources around the ecosystem before you begin the activity. Follow your teacher's instructions to collect as much of the limited resource as possible and answer the following questions as you complete each round of the carrying capacity lab.

Resource Activities	Round 1	Round 2	Round 3	Round 4
Water Collected by Individual				
Water Collected by Class				
Food Collected by Individual				
Food Collected by Class				
Did you survive?				
Percentage of class which survived?				

After Round One

1. Did you successfully collect both resources?

____

2. What was your strategy for collecting the resources?

____

____

____

____

3. What would you change about your performance this round?

4. What percentage of your class survived by collecting both resources?

After Round Two

1. Did you successfully collect both resources?

2. What was your strategy for collecting the resources?

3. What would you change about your performance this round?

4. What percentage of your class survived by collecting both resources? Did this change from the previous round? Why or why not?

After Round Three

1. Did you successfully collect both resources?

2. What was your strategy for collecting the resources?

3. What would you change about your performance this round?

____

____

____

4. What percentage of your class survived by collecting both resources? Did this change from the previous round? Why or why not?

____

____

____

After Round Four

1. Did you successfully collect both resources?

____

2. What was your strategy for collecting the resources?

____

____

____

____

3. What would you change about your performance this round?

____

____

____

4. What percentage of your class survived by collecting both resources? Did this change from the previous round? Why or why not?

____

____

____

Final Question

1. How can limiting factors play a role in animal survival and behavior in an ecosystem?

Name ______________________ Date ______________ Period __________

CHAPTER 7 Global Environmental Issues

Chapter Review

Carefully study the chapter and then answer the following questions.

1. _____ There are 423 national parks in the United States that encompass _____.
   A. more than 84 million acres
   B. less than 48 million acres
   C. more than 100 million acres
   D. less than 28 million acres

2. _____ The land in our national parks covers _____.
   A. 15% of our total acreage
   B. more than 3.6% of our total acreage
   C. less than 3.6% of our total acreage
   D. 2.5% of our total acreage

3. _____ Which of the following are the primary factors that affect land use decisions?
   A. The economic strength of the country
   B. Public and private ownership
   C. Population and land cover
   D. All of the above.

4. _____ The physical land type, such as prairie or desert, is the _____.
   A. land use
   B. land acreage
   C. land cover
   D. All of the above.

5. _____ *True or False?* The manner in which people use the land is the land's inhabitability.

6. _____ Which of the following constitute part of the US land mass?
   A. Around 700,000 square miles of desert
   B. Around 330,000 square miles of Alaska
   C. The land area covered by Hawaii's volcanoes
   D. All of the above.

7. _____ When the use of a resource diminishes the supply, or prevents simultaneous use by others, its use is _____.

A. frivolous
B. nonrivalrous
C. rivalrous
D. justifiable

8. _____ When a resource can be used by one party without denying others access, its use is _____.

A. frivolous
B. nonrivalrous
C. rivalrous
D. justifiable

9. _____ *True or False?* Resources that are privately owned are *not* subject to government regulations.

10. _____ The tendency of an area to grow consistently in size is the area's _____.

A. land strength
B. resource strength
C. economic strength
D. regulatory strength

11. _____ How much of the water on Earth is drinkable?

A. Less than 3%
B. About 6%
C. Between 8.5% and 10%
D. Around 30%

12. _____ Many locations regulate water use and provide a fixed amount of water for a given location or population. These regulations are referred to as water _____.

A. division
B. rationing
C. hoarding
D. sharing

13. _____ The introduction of competition into a market to reduce government control of an energy source is referred to as _____.

A. regulation
B. incorporation
C. deregulation
D. monopolization

14. _____ *True or False?* Monopolization enables customers to select where they want to obtain a utility.

15. _____ *True or False?* Social pressure can provide a positive influence about conservation to the public.

Name ____

16. _____ *True or False?* The economic value of a resource, such as land, may vary among different groups of people.

17. _____ Natural sources of pollution include _____.
   A. human-caused pollution
   B. volcanic eruptions and forest fires
   C. industrial discharge
   D. All of the above.

Vocabulary Review

Match the following terms with the appropriate definition or description.

A. biocapacity
B. ecological footprint
C. green building
D. green chemistry
E. green manufacturing
F. green technology

1. _____ Technology that is intended to reverse or limit the effects of human activity on the environment.

2. _____ The ability of an environment to produce an ongoing supply of resources for a population and to absorb its waste.

3. _____ The design and production of chemical products that have less environmental impact than their traditional counterparts.

4. _____ Structures and the environmentally responsible processes used to design, construct, operate, and maintain these structures.

5. _____ Production processes that have less impact on the environment than their traditional counterparts.

6. _____ The measurement of the impact of human activities on resources ranging from the land we occupy, to the amount of food and other goods we use, to the amount of waste we produce during our lifetime.

Thinking Critically

1. Explain what Albert Einstein meant by the quote: "We cannot solve our problems with the same thinking we used when we created them."

2. What is really the only way to slow or stop climate change? Why does the responsibility belong to humans?

3. How does post-truth, alternative facts, and fake news affect the effort to slow or stop climate change?

Name ______________________________ Date ______________ Period __________

Activity 7A

Eating Local

Objectives

- Identify the commodities produced in your county.
- Locate farmers markets in the local area.
- Explain cooperative programs.
- Improve your research skills.

Materials

- internet access
- activity worksheet

Instructions

Are there locally grown foods in your area? Use the following questions to help you determine where there may be locally produced foods.

Local Commodities

1. Conduct an Internet search to find the top five agricultural commodities produced in your county. The USDA Census of Agriculture website will allow you to search commodities by state and county.
2. List the top five agricultural commodities in your area.

   ____

   ____

   ____

   ____

   ____

3. Did you expect these five to be the top commodities? Explain your answer.

   ____

   ____

   ____

4. Do you know anyone who produces these crops or animals?

   ____

   ____

   ____

Farmers Markets

1. Check the city or county website for your area to see if there are any locally scheduled farmers markets.
2. Are there any farmers markets scheduled in your area?

____

____

3. How often are they held?

____

____

4. Are they open seasonally or year-round?

____

____

Community Gardens

1. Search for community gardens in your area.
2. Are there any community gardens in your area?

____

____

3. Where are they located?

____

____

____

4. What are the requirements for joining the community garden in your area? Are there fees involved?

____

____

____

Cooperatives

1. Search for agricultural community cooperatives in your area. A good search phrase is: "buy locally grown produce in (insert your town/city)."
2. What opportunities exist for buying into a consumer cooperative in your area?

____

____

____

3. What are the fees involved?

____

____

Name ______________________ Date ____________ Period ________

Activity 7B

Foods around the Globe

Objectives

- Determine the staple foods of another country.
- Identify breakfast, snack, and traditional foods of another country.
- Improve research and presentation skills.
- Give a presentation to your class using data you collected.

Materials

- internet access
- library access
- activity worksheet

Instructions

Use the questions below to gather data for a presentation on your assigned/chosen country.

1. Identify your country, its location, and current population.

   A. Country: ______________________

   B. Location: ______________________

   C. Population: ______________________

2. What are the staple foods? (A staple food is a food that provides the majority of calories.)

   ____

   ____

   ____

3. What types of breakfast foods are common? Describe a typical breakfast.

   ____

   ____

   ____

   ____

   ____

4. What are five foods traditionally eaten for lunch and dinner?

   ____

   ____

   ____

   ____

   ____

5. List three popular snack foods.

6. List the most common animal proteins (meat, dairy, eggs) eaten in this country.

7. What food products does this country export to the United States? Do they import food products from the United States? If so, what are the imports?

8. Identify one food product that seems to be very different from those found in traditional American diets.

9. Use the collected information to create a presentation to share with the class.

Name ______________________ Date ______________ Period ________

CHAPTER 8 Population Dynamics

Chapter Review

Carefully study the chapter and then answer the following questions.

1. _____ The field of study that researches the size and composition of a population as it relates to its environment is called population _____.
   A. arrangement
   B. dispersion
   C. dynamics
   D. capacity

2. _____ The number of organisms of a given species living in a given area is _____.
   A. its maximum capacity
   B. a population
   C. its arrangement
   D. All of the above.

3. _____ *True or False?* If the arrangement of resources is not properly aligned, survival of some species may be at risk.

4. _____ *True or False?* Too much space between individuals in a rural area may cause stress for some people.

5. _____ *True or False?* Building vertically restricts or reduces the number of people that can live in an urban area.

6. _____ Which of the following is performed nationwide to gather information about every member of a population?
   A. Estimate
   B. Survey
   C. Census
   D. All of the above.

7. _____ The process of using a group of subjects that is representative of the entire population is population _____.

A. surveying

B. representation

C. estimation

D. sampling

8. _____ Which of the following methods is used to estimate the size of a population where it is impractical to count every individual?

A. Mark-recapture

B. Mark-estimation

C. Mark-sample

D. Mark-survey

Use the following formula to answer questions 9 and 10.

Population estimate = Total number in second sample × Number marked and released in first sample ÷ total marked number found in second sample

9. _____ The forest service finds 120 white-tailed deer in a first sampling and marks all 120 before releasing them. In the second sampling, the service counts 101 deer. Eighty-five of the deer in the second sampling had tags from the first sampling. What is the population estimate?

A. 85

B. 120

C. 132.5

D. 142.5

10. _____ The forest service finds 85 red-tail foxes in a first sampling and marks all 85 before releasing them. In the second sampling, the service counts 80 foxes. Seventy-five of the foxes in the second sampling had tags from the first sampling. What is the population estimate?

A. 80

B. 86

C. 91

D. 99

11. _____ The method of population estimation that uses a small sample of a population taken in a predetermined area is _____.

A. recapture sampling

B. quadrat sampling

C. population sampling

D. All of the above.

Name ______________________________

12. _____ Researchers are working on a project to maintain populations of native pollinators. They need to determine whether a meadow contains enough pollen-producing clover plants to support the pollinators. The researchers place thirty $2' \times 2'$ quadrats randomly in the meadow and determine the presence or absence of clover within each quadrat. Note that at least 2% of the total area should be sampled to acquire an accurate estimate. How large is the meadow if the quadrats cover 3% of the total area?

A. 2000 ft^2

B. 4000 ft^2

C. 6000 ft^2

D. 8000 ft^2

13. _____ Of the 30 quadrats placed, 20 held at least one clover specimen. Based on this information, what percent of the total meadow contains clover?

A. 3%

B. 20%

C. 67%

D. 80%

14. _____ Population ecology is the study of how population sizes and groupings change over time. Which of the following factors are considered when studying population ecology?

A. Population stability

B. Migration patterns

C. Limiting factors

D. All of the above.

15. _____ Which of the following would likely be a limiting factor for a herd of zebras?

A. Drought

B. Abundant grass

C. Shade from trees

D. Open plains

16. _____ *True or False?* A stable population remains steady in number.

17. _____ The movement of a being from one region or climate to another is _____.

A. emigration

B. immigration

C. litigation

D. elimination

18. _____ Which of the following occurs when members of a species leave one area and move to another to meet their individual needs?

A. Immigration

B. Litigation

C. Elimination

D. Emigration

19. _____ Which of the following is a spacing pattern of individuals in a given habitat and their proximity to other individuals of the same species?

A. Carrying dispersion
B. Immigration pattern
C. Random positioning
D. Population dispersion

20. _____ The type of population dispersion that provides the most possible space between individuals in a given location is _____.

A. uniform dispersion
B. clumped dispersion
C. random dispersion
D. elaborate dispersion

21. _____ Which of the following is a type of population dispersion in which individuals are spaced at unpredictable distances from one another?

A. Uniform dispersion
B. Clumped dispersion
C. Random dispersion
D. Elaborate dispersion

22. _____ Which of the following is the type of population dispersion in which individuals live in clusters with significant space between other groups of the same species?

A. Uniform dispersion
B. Clumped dispersion
C. Random dispersion
D. Elaborate dispersion

23. _____ A plant's excretion of toxins to inhibit the growth of other plants is referred to as _____.

A. uniformity
B. aggression
C. dispersion
D. alleopathy

24. _____ The number of living creatures in relationship to the space available, or a measurement of the number of individuals per unit area or volume is population _____.

A. capacity
B. density
C. alleopathy
D. dispersion

25. _____ *True or False?* Arrangement of resources is not a limiting factor for most animals.

Name ______________________________

26. _____ The number of living organisms that a habitat can support without degradation is the habitat's _____.

A. carrying capacity

B. population dispersion

C. home capacity

D. territory limit

27. _____ When an individual within a population shows hostility toward another individual, its behavior is _____.

A. specific

B. aggressive

C. limiting

D. strategic

28. _____ The marking and physical defense of a home range is referred to as _____.

A. exclusion

B. aggression

C. territorialism

D. All of the above.

29. _____ *True or False?* Intraspecific territoriality is between different species.

30. _____ *True or False?* Interspecific territoriality is between members of the same species.

31. _____ Which of the following occurs when the population of a species is too great for its environment?

A. More prone to disease

B. Increased vulnerability to predators

C. Starvation

D. All of the above.

32. _____ *True or False?* Thomas Malthus believed that the growth of a population would always outreach its ability to feed itself.

Thinking Critically

1. Why was Thomas Malthus considered an extremist?

2. What are the limiting factors of the human population?

3. Thomas Malthus discussed the belief that civilization would never reach a point of true prosperity and that there would never come a time when all people could live harmoniously. Do you agree or disagree? Explain your answer.

Name ______________________ Date ______________ Period __________

Activity 8A

Population Dynamics

Part 1: Mule Deer

Objectives

- Observe how and why population sizes change.
- Calculate the changes in several animal populations.

Materials

activity worksheet

Instructions

For each problem, calculate what the animal population will be in five years. You must perform the calculations for each year. In all populations, assume the initial number and all births are half male and half female. If you have an odd number of animals, the male population has the additional member. Review the example.

Example: There are 100 mule deer in an area. The yearly birth rate is one fawn per doe. Predators will take 25% of the population, and hunters will kill 25% of the population. What will the population of this herd be in five years?

Year 1			
Item	**Change**	**Total**	**Notes**
Beginning population	—	100	There are 50 males and 50 females
Births	+50	150	50 does at 1 fawn per doe
Predator loss	–38	112	25% of 150 = 37.5 (you cannot kill 0.5 of an animal and must round this number up.)
Hunter loss	–38	74	25% of 150 = 37.5
Year 2			
Item	**Change**	**Total**	**Notes**
Beginning population	—	74	37 males and 37 females
Births	+37	111	37 does at 1 fawn per doe
Predator loss	–28	83	25% of 111 = 28
Hunter loss	–28	55	25% of 111 = 28
Year 3			
Item	**Change**	**Total**	**Notes**
Beginning population	—	55	28 males and 27 females
Births	+27	82	27 does at 1 fawn per doe
Predator loss	–21	61	25% of 82 = 21
Hunter loss	–21	40	25% of 82 = 21
Year 4			
Item	**Change**	**Total**	**Notes**
Beginning population	—	40	20 males and 20 females
Births	+20	60	20 does at 1 fawn per doe
Predator loss	–15	45	25% of 60 = 15
Hunter loss	–15	30	25% of 60 = 15

Continued

Year 5			
Item	**Change**	**Total**	**Notes**
Beginning population	—	30	15 males and 15 females
Births	+15	45	15 does at 1 fawn per doe
Predator loss	–12	33	25% of 45 = 12
Hunter loss	–12	21	25% of 45 = 12

Total population after five years: 21

1. What would you do to manage this mule deer population more effectively?

____

____

____

____

Part 2: Rocky Mountain Elk

1. This data is for a herd of 500 Rocky Mountain elk. Their yearly birth rate is one calf for every cow. Predators will take 10% of the population each year. There is no hunting in this area. What will the population be in five years?

Year 1			
Item	**Change**	**Total**	**Notes**
Beginning population	—		
Births			
Predator loss			
Hunter loss			
Year 2			
Item	**Change**	**Total**	**Notes**
Beginning population	—		
Births			
Predator loss			
Hunter loss			
Year 3			
Item	**Change**	**Total**	**Notes**
Beginning population	—		
Births			
Predator loss			
Hunter loss			

Continued

Name ______________________________

Year 4			
Item	**Change**	**Total**	**Notes**
Beginning population	—		
Births			
Predator loss			
Hunter loss			
Year 5			
Item	**Change**	**Total**	**Notes**
Beginning population	—		
Births			
Predator loss			
Hunter loss			

Total population after five years: ______________________________

2. What would you do to manage this Rocky Mountain elk population more effectively?

____

____

____

____

Part 3: Bison

1. This data is for a herd of 200 bison. Their yearly birth rate is one calf for every two cows. They will lose 10% of the population to hunters. Predators do not take any animals in this range. What will the population be in five years?

Year 1			
Item	**Change**	**Total**	**Notes**
Beginning population	—		
Births			
Predator loss			
Hunter loss			
Year 2			
Item	**Change**	**Total**	**Notes**
Beginning population	—		
Births			
Predator loss			
Hunter loss			

Continued

Year 3			
Item	Change	Total	Notes
Beginning population	—		
Births			
Predator loss			
Hunter loss			

Year 4			
Item	Change	Total	Notes
Beginning population	—		
Births			
Predator loss			
Hunter loss			

Year 5			
Item	Change	Total	Notes
Beginning population	—		
Births			
Predator loss			
Hunter loss			

Total population after five years: ______________________________

2. What would you do to manage this bison population more effectively?

____

____

____

____

Part 4: Ducks

1. This data is for a flock of 1000 ducks. Their yearly birth rate will be five ducklings per hen. They will lose 25% of the population to hunters, and 25% of the population to predators. What will the population be in five years?

Year 1			
Item	Change	Total	Notes
Beginning population	—		
Births			
Predator loss			
Hunter loss			

Continued

Name ________________________________

Year 2			
Item	**Change**	**Total**	**Notes**
Beginning population	—		
Births			
Predator loss			
Hunter loss			

Year 3			
Item	**Change**	**Total**	**Notes**
Beginning population	—		
Births			
Predator loss			
Hunter loss			

Year 4			
Item	**Change**	**Total**	**Notes**
Beginning population	—		
Births			
Predator loss			
Hunter loss			

Year 5			
Item	**Change**	**Total**	**Notes**
Beginning population	—		
Births			
Predator loss			
Hunter loss			

Total population after five years: ________________________________

2. What would you do to manage this duck population more effectively?

____

____

____

____

Part 5: Quail

1. This data is for a group of 1000 bobwhite quail. Their yearly birth rate will be ten chicks per hen. They will lose 50% of their population to predators, and 25% population to hunters. What will the population be in five years?

Year 1			
Item	**Change**	**Total**	**Notes**
Beginning population	—		
Births			
Predator loss			
Hunter loss			

Year 2			
Item	**Change**	**Total**	**Notes**
Beginning population	—		
Births			
Predator loss			
Hunter loss			

Year 3			
Item	**Change**	**Total**	**Notes**
Beginning population	—		
Births			
Predator loss			
Hunter loss			

Year 4			
Item	**Change**	**Total**	**Notes**
Beginning population	—		
Births			
Predator loss			
Hunter loss			

Year 5			
Item	**Change**	**Total**	**Notes**
Beginning population	—		
Births			
Predator loss			
Hunter loss			

Total population after five years: ____________________

1. What would you do to manage this quail population more effectively?

____

____

____

Name ______________________________ Date ______________ Period __________

Activity 8B

Populations and Survival Needs

Objectives

- To understand the importance of a suitable habitat for an animal population.
- To identify the essential components of habitat.
- To identify the factors that affect wildlife populations in changing ecosystems.
- To understand that a population will increase in size until limiting factors cause fluctuations in the population.
- List examples of limiting factors.

Background

In this activity, you and your classmates will represent a population of deer and the basic necessities that the deer need to survive. For the purpose of this activity, we will assume that the deer have enough space in which to live.

Questions to Consider

1. What are the fundamental necessities for an animal to survive?

____

____

____

2. What factors affect the ability of wildlife to successfully reproduce and maintain their population?

____

____

____

Materials

- large indoor or outdoor area
- easel with paper or white board
- markers/dry erase markers
- spray paint/chalk/tape for marking boundaries
- tablets/clipboards
- pencil
- data record sheets
- 64 blue cards
- 64 brown cards
- 64 green cards

Set Up the Game

1. Draw parallel lines on the ground 10 to 20 yards apart.
2. Draw a copy of the activity chart on the paper or white board.
3. Count off in fours. All students who are ones line up behind one of the parallel lines with their backs toward the line. These students are the deer.
4. The remaining students, the twos, threes, and fours, should form three lines behind the opposite line with their backs facing the deer. These students represent the basic necessities of food, water, and shelter.
5. Distribute one of each color card to each of the students.

Instructions for the Deer

6. You will use the cards and hand gestures to indicate the resource that you are seeking. **Note:** You may choose any one of your needs during each round but you may *not* change what you are looking for until the beginning of the next round.
   - When looking for food, hold a green card and clamp your hand over your stomach.
   - When looking for water, hold a blue card and put your hand over your mouth.
   - When looking for shelter, hold a brown card and put your hand(s) over your head.

Instructions for the Resources

7. You will use the cards and hand gestures to indicate the resource that you represent. **Note:** You may choose to be any one of the resources during each round, but you must remain the same resource until the beginning of the next round.
   - Hold a green card and put your hand over your stomach to indicate that you are food.
   - Hold a blue card and put your hand over your mouth to indicate that you are water.
   - Hold a brown card and put your hand(s) over your head to indicate that you are shelter.
8. You are to remain in place until a deer comes to get you.

Play the Game

9. All players should be lined up with their backs to the students behind the other line.
10. Select your card and make the gesture that represents the resource. The deer choose what resource they are seeking and the other students determine which resource they represent.
11. Your instructor will count to three, and at the count of three everyone turns to face the other group, continuing to hold the cards and gestures.
12. Each of the deer walks quickly to the resource they need. **Note:** Resources do not move and must wait until a deer comes to get them.
13. Each deer that reaches the resource it is seeking, takes the resource (student representing the resource) to the deer side of the habitat. This represents the deer successfully meeting its needs. By meeting its needs, the deer reproduces and each of the resources used by the deer become deer in the next round.
14. Any deer that fails to find its food, water, or shelter dies and becomes part of the habitat and will be a resource in the next round. **Note:** If two or more deer approach the same resource, the deer that arrives first survives and the other deer dies.
15. Play 12 rounds of the game and record the number of deer alive at the beginning and end of each round on the display chart. Copy the numbers from the whiteboard in the following chart.

Name ______________________________

Round	# of deer at beginning of round	# of deer at end of round
0		
1		
2		
3		
4		
5		
6		
7		
8		
9		
10		
11		
12		

Round	# of deer at beginning of round	# of deer at end of round
0		
1		
2		
3		
4		
5		
6		
7		
8		
9		
10		
11		
12		

Round	# of deer at beginning of round	# of deer at end of round
0		
1		
2		
3		
4		
5		
6		
7		
8		
9		
10		
11		
12		

Discuss the Results

1. What did you notice about the deer population as we played the game? For example, how and when did the size of the population change?

____

____

____

____

2. Use the data from the activity chart to graph your results.

NOTE Each round represents one year.

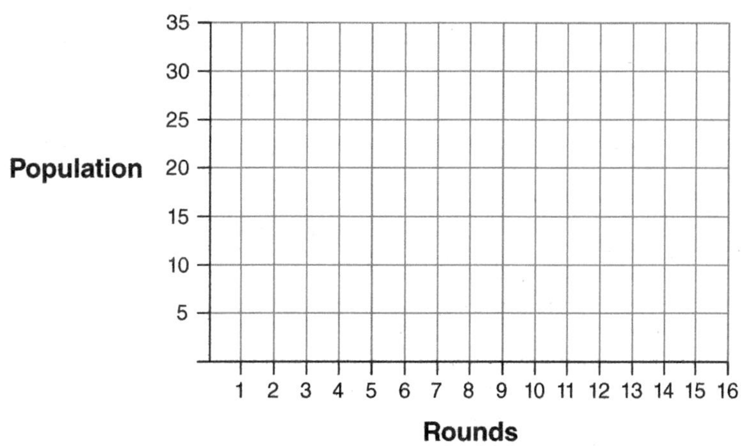

Goodheart-Willcox Publisher

Questions for Review

1. Why is a suitable habitat important for an animal population?

2. What are the essential components of a habitat? How do these differ between species?

3. What factors of a changing ecosystem affect a wildlife population?

4. Give examples of limiting factors that would affect a deer population.

Name ______________________ Date ____________ Period ________

Activity 8C

World Population Growth

Objectives

- Conceptualize the speed of the world's population growth.
- Graph population data for visual of population growth.
- Discuss growth predictions.

Instructions

According to the United Nations Population Division, roughly 83 million people are added to the world population each year, and the upward trend in population size is expected to continue. Consider what this means for our planet. Use the following data to graph the world population growth from 1000 CE to the projected growth in 2100 CE. Use a dotted line to indicate the projected growth.

Year	World Population (Millions)
1000	275
1200	360
1400	350
1600	500
1760	770
Year	**World Population (Billions)**
1804	1.0
1900	1.6
1927	2.0
1952	2.6
1960	3.0
1970	3.7
1975	4.0
1980	4.5
1990	5.3
2000	6.1
2010	6.9
2015	7.4
2020	7.8
2023	8.0
2030	8.5*
2050	9.7*
2100	10.3*
*Estimates by 2020 United Nations Department of Economic and Social Affairs Report	

Goodheart-Willcox Publisher

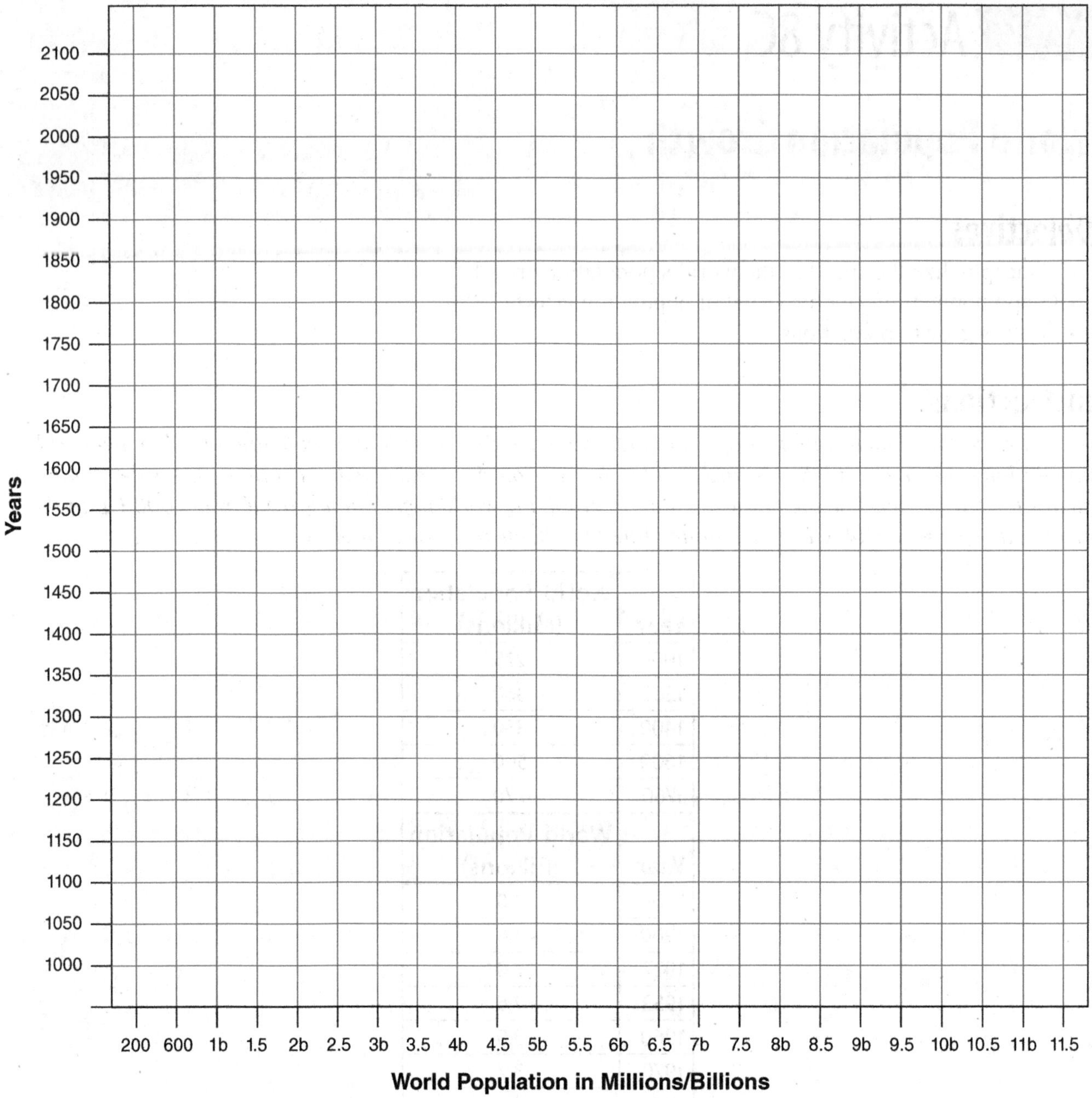

Goodheart-Willcox Publisher

Name ____________________

Questions to Consider

1. Is the projected growth rate accurate? Why or why not?

2. Which continent is predicted to double its population by the year 2050?

3. How many countries are predicted to have a decrease in their population? Why do you think this is the case?

4. Which countries have the highest growth rate, developed or developing countries? Why do you think these countries have the highest growth rate?

5. How many people have ever lived?

6. What is demographic transition? How does it affect population growth?

Notes

Name ______________________________ Date ______________ Period __________

CHAPTER 9 What Is Soil?

Chapter Review

Carefully study the chapter and then answer the following questions.

1. _____ The carbon-based materials found in soil are _____.
   A. decomposed plants
   B. decomposed animals
   C. organic matter
   D. All of the above.

2. _____ *True or False?* Soil that contains much organic matter does not hold water well.

3. _____ Which of the following is the ability of soil to hold water molecules for plant growth?
   A. Drainage capacity
   B. Water-holding capacity
   C. Flow capacity
   D. Nutrient capacity

4. _____ *True or False?* The bulk of soil is composed of sand, silt, and clay particles.

5. _____ Sandy soils _____.
   A. tend to be acidic
   B. are low in nutrients
   C. are light and dry
   D. All of the above.

6. _____ Clay soils _____.
   A. are light and dry
   B. have good drainage
   C. remain wet and cold in winter
   D. All of the above.

7. _____ Silty soils _____.
   A. retain moisture and have a high fertility rating
   B. are dry, light, and low in nutrients
   C. hold water tightly and remain wet
   D. All of the above.

8. _____ Which of the following feel similar to grains of table salt and are the largest soil particles?

A. Silt particles
B. Clay particles
C. Sand particles
D. Dirt particles

9. _____ *True or False?* The pore spaces in compacted soil are large open spaces.

10. _____ Which of the following soil particles are microscopic, feel like baking flour, and create large pore spaces?

A. Silt particles
B. Clay particles
C. Sand particles
D. Dirt particles

11. _____ Which of the following soil particles are plate-shaped, create small pore spaces, and cause the soil to remain wet once it is saturated?

A. Silt particles
B. Clay particles
C. Sand particles
D. Dirt particles

12. _____ The geological material from which soil is formed is the soil's _____.

A. bedrock material
B. organic material
C. parent material
D. All of the above.

13. _____ Which of the following contribute to the weathering of parent material?

A. Wind and water erosion
B. Decomposition of biological organisms
C. Temperature changes
D. All of the above.

14. _____ The surface horizon, which is just below the thin organic layer of soil, is the _____.

A. R-horizon
B. B-horizon
C. C-horizon
D. A-horizon

15. _____ The thin, organic layer of soil just immediately below the vegetation is the _____.

A. A-horizon
B. O-horizon
C. B-horizon
D. C-horizon

Name ______________________________

16. _____ The parent material or substratum is the _____.

A. A-horizon

B. O-horizon

C. C-horizon

D. R-horizon

17. _____ The bedrock over which the soil profile is laid is the _____.

A. R-horizon

B. A-horizon

C. O-horizon

D. C-horizon

18. _____ Tends to be rockier and denser than the other layers and is referred to as the _____.

A. A-horizon

B. B-horizon

C. C-horizon

D. R-horizon

19. _____ *True or False?* Soils that lack significant levels of organic matter are mineral soils.

20. _____ Organic soils have about _____.

A. 5% organic material

B. 10% organic material

C. 15% organic material

D. 20% organic material

21. _____ Clumps of soil aggregates are called _____.

A. particles

B. portions

C. peds

D. plates

22. _____ The measure of a soil's compaction is its _____.

A. density

B. profile

C. structure

D. peds

23. _____ *True or False?* The size of a soil's pores determine how freely water will flow through the horizons.

24. _____ Which of the following apply to gray soil?

A. Good drainage

B. Good porosity

C. Lack of oxygen and iron oxidation

D. All of the above.

25. _____ *True or False?* One of the most important aspects of soil infiltration is that it must allow the soil to drain quickly and not hold water.

26. _____ *True or False?* When soil is the driest, it has the least infiltration potential.

27. _____ When all pore spaces are filled with water and can no longer support infiltration, the soil is _____.

A. saturated
B. infiltrated
C. percolated
D. excavated

28. _____ The movement of water within the soil profile is due primarily to gravity pulling the water downward. This movement is called _____.

A. saturation
B. infiltration
C. percolation
D. excavation

29. _____ *True or False?* A plant may not be able to absorb the nutrients it needs if the soil's pH is not in the proper range.

30. _____ *True or False?* There must be an adequate cation exchange capacity for plants to absorb nutrients.

31. _____ *True or False?* Land capability classes I, II, III, and IV are suitable for cropping, pasture, rangeland, or forests.

32. _____ *True or False?* Land capability classes are suitable for cropping or pasture.

For questions 33–36, match the LLC land subclass with the appropriate designation.

A. c
B. e
C. s
D. w

33. _____ A designation that indicates the soil is susceptible to erosion.

34. _____ A designation that means that the soil has excessive wetness or poor drainage.

35. _____ The designation that indicates climatic limitations, such as arid areas or those with excessive rainfall.

36. _____ The designation that indicates limitations with the rooting zone.

Name ____

37. _____ A layer of poorly drained, hard soil through which roots cannot permeate is a _____.
A. fragipan layer
B. permeated layer
C. drainage layer
D. capability layer

38. _____ A layer of sand and gravel, glacial till, or bedrock that greatly limits rooting through the soil horizon is a _____.
A. classified layer
B. limiting layer
C. drainage layer
D. capability layer

Vocabulary Review

Match the soil nutrients to the appropriate description.

A. nonmineral nutrients
B. primary macronutrients
C. secondary macronutrients
D. micronutrients

1. _____ Boron, copper, chlorine, iron, manganese, molybdenum, nickel, and zinc
2. _____ Carbon, hydrogen, and oxygen
3. _____ Calcium, magnesium, and sulfur
4. _____ Nitrogen, phosphorus, and potassium

Match the soil horizons to the corresponding content.

5. _____ Unweathered parent material
6. _____ Organic matter
7. _____ Mixture of sand, silt, and clay
8. _____ Organics mixed with mineral matter
9. _____ Parent rock

Designua/Shutterstock.com

Match the physical properties of soil to the appropriate definition or description.

A. soil classification
B. soil horizon
C. soil infiltration
D. soil porosity
E. soil profile
F. soil reactivity
G. soil structure
H. soil survey
I. soil texture

10. _____ The process of classifying soil types and other soil properties in a given area.

11. _____ The amount of sand, silt, and clay present in a given soil.

12. _____ The soil's ability to allow water movement *into* and *through* the soil profile.

13. _____ The clumping or binding of soil particles into aggregates.

14. _____ A distinct layer of soil.

15. _____ The arrangement of soil layers (horizons).

16. _____ The separation of soil into groups each having similar characteristics and potentially similar behavior.

17. _____ A measure of the active hydrogen ion (H^+) concentration of a substance.

18. _____ A measure of the voids or pore spaces in soil.

Match the soil pH to the appropriate soil.

A. acidic soils
B. alkaline soils
C. neutral soils

19. _____ Soils with a pH that falls within the range of 7 to 14

20. _____ Soils with a pH of exactly 7

21. _____ Soils with a pH that falls within the range of 0 to 7

Match each of the soil classifications to the appropriate description.

A. aeolian soils
B. alluvial soils
C. colluvial soils
D. cumulose soils
E. lacustrine soils
F. loess soils
G. marine soils

22. _____ Soils that have been moved by wind

23. _____ Soils moved and deposited in a new location by flowing water

24. _____ Soils that settle in sedimentary deposits of lakes

25. _____ that were once the beds of ancient seas

26. _____ Soil formed from parent material that is not moved but formed by the deposition of organic material.

27. _____ Soils moved by gravity, such as when soils fall from the sides of mountains

Name ____

Thinking Critically

1. How can we ensure our soil is conserved for future generations?

2. Can an increase in hydroponic operations reduce the number of land operations without reducing food and fiber production? Explain your answer.

3. How does soil affect the development of buildings?

Notes

Name ______________________ Date ____________ Period ________

Activity 9A

Edible Soil Profile

Objectives

- Identify the soil horizons.
- Explain the function of each soil horizon.
- Improve your research skills.

Materials

- internet access
- activity worksheet
- clear plastic cup
- one chocolate sandwich cookie
- crumbled chocolate chip cookies
- vanilla pudding
- chocolate pudding
- gummy worms
- sprinkles
- spoons and/or spatulas

Section 1: Soil Formation

1. List the five factors that influence soil formation.

____

____

____

____

____

2. Which factor is most important in soil formation? Explain your answer.

____

____

____

____

____

____

____

____

Section 2: Create Your Soil Horizon

Follow the instructions to create a soil horizon model and then answer the questions about your model. Wash your hands before you begin. Do not *eat the profile until given permission to do so.*

1. Place the sandwich cookie (horizon R) in the bottom of a plastic cup.
2. Place a layer of crumbled chocolate chip cookies (horizon C) on top of the sandwich cookie.
3. Place a layer of vanilla pudding (horizon B) on top of the crumbled chocolate chip cookies.
4. Place a layer of chocolate pudding (horizon A) on top of the vanilla pudding.
5. Add sprinkles (horizon O) and place 2–3 gummy worms on the top.

Horizon Functions

Detail the function of each listed horizon in soil.

1. **Horizon O** (sprinkles/worms)

2. **Horizon A** (chocolate pudding)

3. **Horizon B** (vanilla pudding)

4. **Horizon C** (crumbled cookies)

5. **Horizon R** (sandwich cookie)

Name ____

Section 3: Analysis and Conclusion

1. Draw, color, and label the horizons of your edible soil.

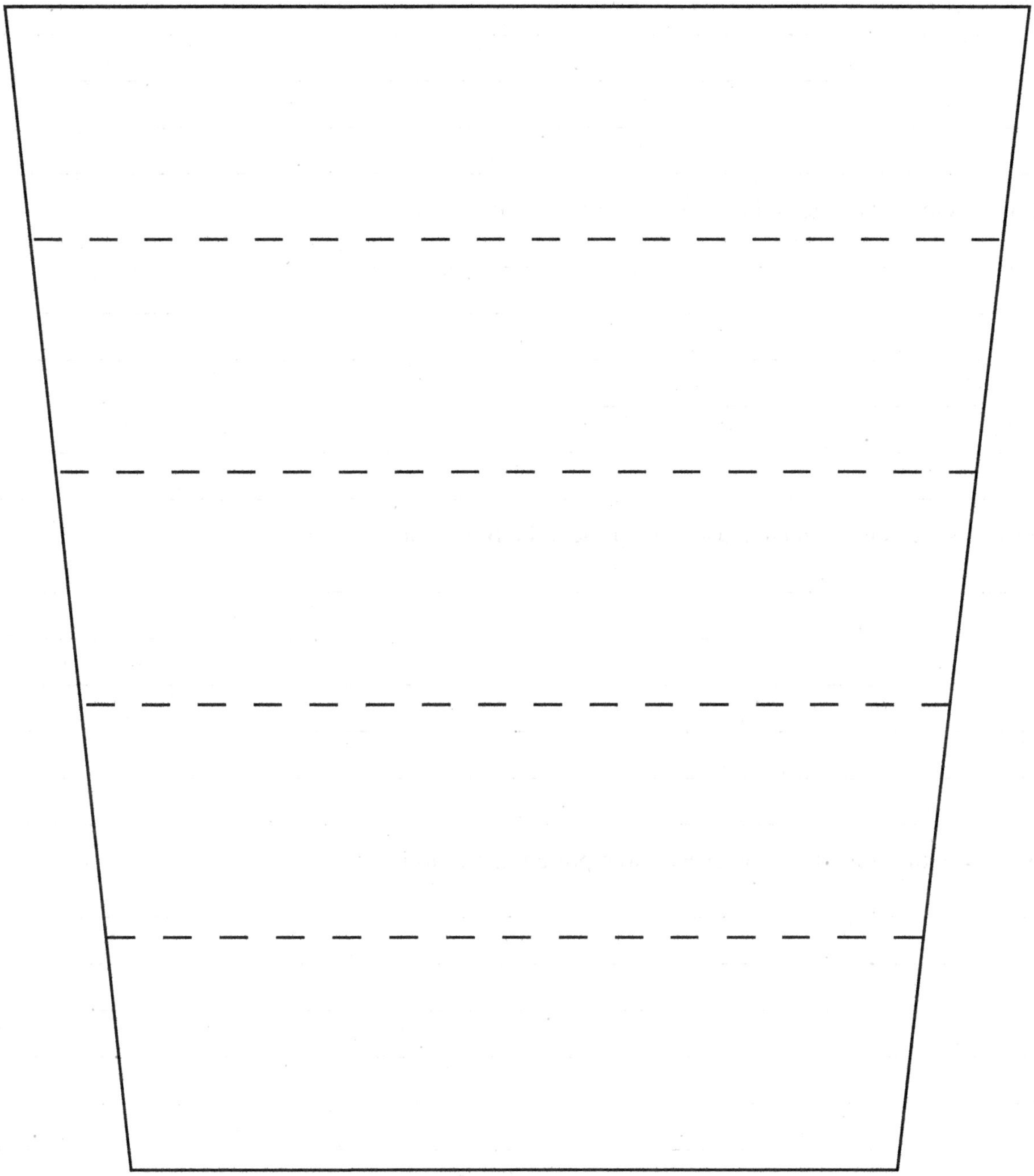

Goodheart-Willcox Publisher

2. Why were the sprinkles and worms used to represent the O horizon?

3. Why was the chocolate pudding used to represent the A horizon?

4. Why was the vanilla pudding used to represent the B horizon?

5. Why were the crumbled cookies used to represent the C horizon?

6. Why was the sandwich cookie used to represent the H horizon?

Name ______________________ Date ____________ Period ________

Activity 9B

Comparing Soil Particles

Objectives

- Visualize the difference in size of soil particles.
- Calculate percentages.
- Use a soil matrix.

Materials

- activity sheet
- cereal and/or candy (3 types)
- paper towels
- plastic cups

Particle Sizes

Soil particles range in size by their composition.

- Clay particles are the smallest (less than 0.002 mm)
- Silt particles are larger than clay particles but smaller than sand particles (0.002 mm to 0.05 mm)
- Sand particles are the largest soil particles (0.05 mm to 2.0 mm)

Instructions

Divide the class into groups of 2–3 students. Each group will be given 3 plastic cups (labeled A, B, and C) with a prepared "soil" made of cereal and/or candy. You will determine which cereal/candy represents each type of particle and use the soil triangle to determine what type of soil is in each cup.

1. Pour the contents of your cup on a paper towel and spread the pieces to enable accurate counting.
2. Determine which candy/cereal represents each type of soil particle. Write the name of each material in the appropriate space.

   Clay: ____________________

   Silt: ____________________

   Sand: ____________________
3. Count each particle type.

   Clay: A ____________ B ____________ C ____________

   Silt: A ____________ B ____________ C ____________

   Sand: A ____________ B ____________ C ____________

   Total: A ____________ B ____________ C ____________
4. Calculate the percentage of each type of particle.

   Clay: A ____________ B ____________ C ____________%

   Silt: A ____________ B ____________ C ____________%

   Sand: A ____________ B ____________ C ____________%

5. Use the soil triangle to determine the type of "soil" in each cup.

Soil type in cup A: ___________________________

Soil type in cup B: ___________________________

Soil type in cup C: ___________________________

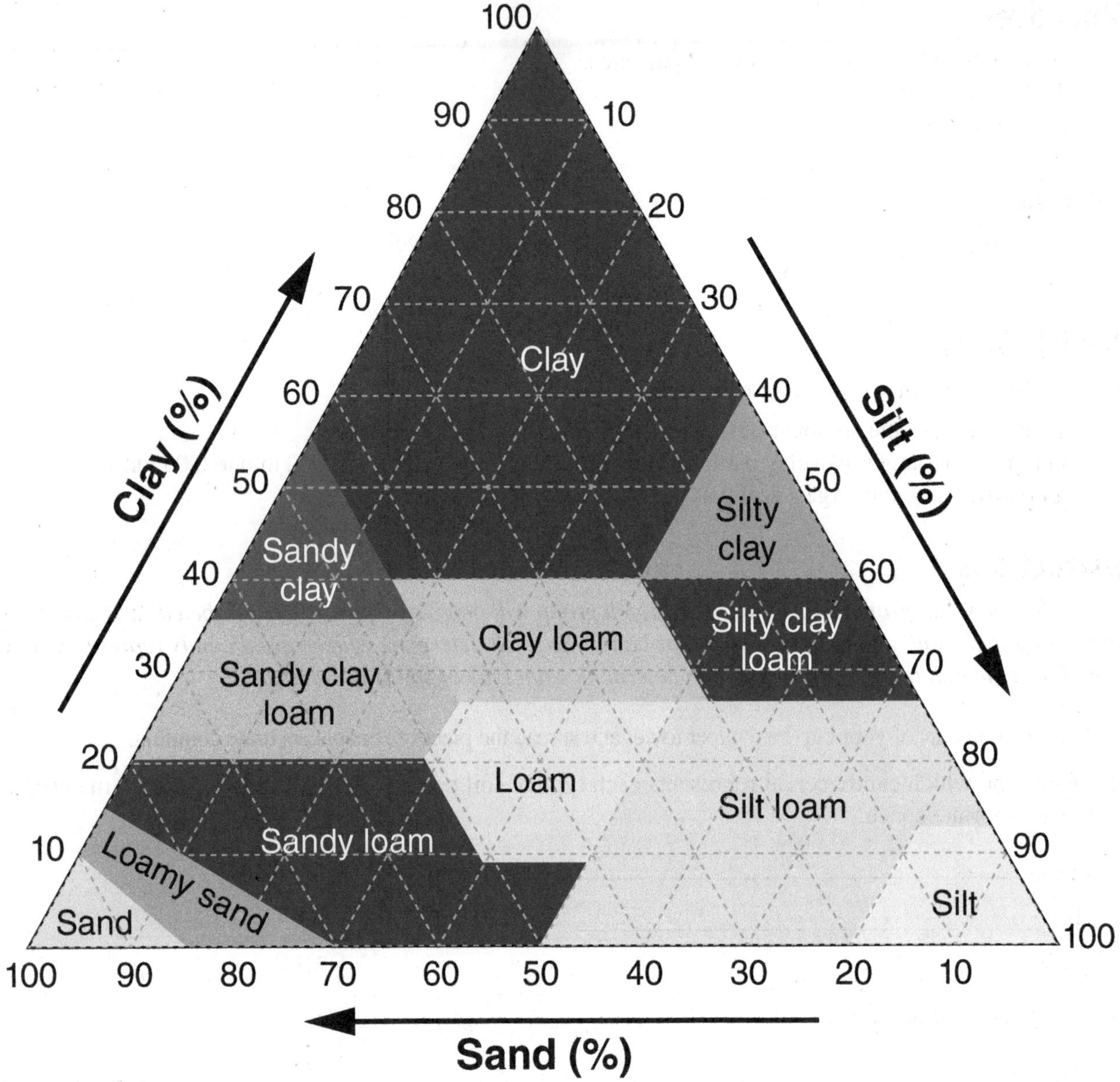

Becris/Shutterstock.com

Name ______________________ Date ______________ Period __________

CHAPTER 10 Soil Erosion, Leaching, and Pollution

Chapter Review

Carefully study the chapter and then answer the following questions.

1. _____ *True or False?* Degraded soil is often fully restored to its original productivity with time and other resources.

2. _____ *True or False?* Soil compaction is the most common form of soil degradation.

3. _____ Which of the following is *not* true about soil compaction?
   A. Clayey soils compact more densely than sandy soils
   B. Wet soils compact more densely than dry soils
   C. Sandy soils compact more densely than clayey soils
   D. Plant roots cannot grow through severely compacted soils

4. _____ *True or False?* The removal *and* displacement of the top layer of soil is soil erosion.

5. _____ *True or False?* Soil that is dislodged and moved to a new location has the same level of productivity as it had in its natural state.

6. _____ *True or False?* Wind erosion is more common on most land than water erosion.

7. _____ Which of the following causes water to remain at or near the surface of the soil?
   A. Soil saturation
   B. Soil compaction
   C. Slow infiltration
   D. All of the above.

8. _____ *True or False?* Splash erosion moves soil particles a significant distance.

9. _____ What is the maximum rate of annual soil loss that will permit sustainable productivity?
   A. T value
   B. E value
   C. Tolerable value
   D. Sustainable value

10. _____ Which of the following is often the only viable solution to rill erosion?

A. Normal soil tillage

B. Conservation structures

C. Subsurface drainage

D. Sediment removal

11. _____ Which of the following is often required to remedy gully erosion?

A. Normal soil tillage

B. Conservation structures

C. Subsurface drainage

D. Sediment removal

12. _____ The sensitivity of soil to the effects of wind and water on soil structure is indicated by which of the following?

A. Gully index (GI)

B. T index (TI)

C. Erosion index (EI)

D. Erodibility index (EI)

13. _____ Which of the following removes soil from the bank of a pond or lake?

A. Shoreline erosion

B. Channel erosion

C. Streambank erosion

D. All of the above.

14. _____ Which of the following occurs when soil from the streambank erodes and deposits in the channel of a stream?

A. Shoreline erosion

B. Channel erosion

C. Streambank erosion

D. All of the above.

15. _____ Which of the following is the removal of riverbanks or beds that occurs naturally as water flows and collects some sediment?

A. Shoreline erosion

B. Channel erosion

C. Streambank erosion

D. All of the above.

16. _____ Which of the following is an example of soil structure that leads to erosion?

A. Soils that are low in organic matter

B. Soils that are severely compacted

C. Soils that have high sand content

D. All of the above.

Name ____

17. _____ Which of the following increases the probability of soil erosion?

A. The steepness and length of the slope

B. The removal of vegetative cover

C. Weak soil structure and lack of organic matter

D. All of the above.

18. _____ What occurs when soil is repeatedly saturated with excess amounts of rain or irrigation?

A. The roots store increased amounts of minerals

B. Plant nutrients float on the soil's surface

C. Plant nutrients flow below the root zone

D. All of the above.

19. _____ *True or False?* Leaching contributes to water pollution.

20. _____ What is *eutrophication*?

A. The rapid death of aquatic plant life

B. The accumulation of dissolved oxygen

C. The healthiest state of a body of water

D. The rapid growth of aquatic plant life

21. _____ Which of the following soil pollutants enter the soil primarily through the application of sewage sludge and controlled release fertilizers?

A. Cleaning agents

B. Plastics

C. Deicing materials

D. All of the above.

22. _____ *True or False?* Soil and water contaminated with petroleum products will not support plant or animal life.

23. _____ *True or False?* Very small amounts of heavy metals, such as lead, arsenic, or mercury, contaminate soil to a level that is toxic to plants, animals, and humans.

24. _____ Which of the following consumer products contaminate soil and water?

A. Prescribed/over-the-counter medications

B. Laptop and remote control batteries

C. Cleaning agents, soaps, and bleach

D. All of the above.

Vocabulary Review

Match each of the stages of water erosion to the most appropriate description.

A. gully erosion
B. particle erosion
C. rill erosion
D. sheet erosion
E. splash erosion

1. _____ Occurs when water carries soil particles and carves out small, uniform, and predominantly straight channels.

2. _____ The breaking apart of soil aggregates due to the force of raindrops on bare soil.

3. _____ Occurs when rills combine and channel water and dislodged soil particles into larger channels.

4. _____ Occurs when water accumulates on the soil's surface, pools, and flows over the surface.

Match the following terms with the appropriate definition or description.

A. shoreline erosion
B. channel erosion
C. streambank erosion
D. slope
E. leaching
F. eutrophication
G. soil compaction
H. soil degradation

5. _____ A process by which a body of water becomes enriched with excess nutrients.

6. _____ The rise or fall of the land.

7. _____ The removal of stream and riverbanks or beds that occurs naturally as water flows and collects some sediment.

8. _____ Occurs when soil from the streambank erodes and deposits in the channel of the stream.

9. _____ Erosion that removes soil from the bank of a pond or lake.

10. _____ The loss of soil minerals (nutrients) due to excessive rain or irrigation.

11. _____ The reduction of soil productivity caused by human or natural causes.

Name ____

Thinking Critically

1. Soil erosion is a natural process. Why is it presently of great concern? Explain how vegetative cover protects soil from erosion.

2. How has overburden from mining operations polluted soil and water?

3. Why would running water be considered more powerful than waves, gravity, wind, and glaciers in terms of soil erosion?

4. What effect might increased soil erosion have on global climate change?

5. Should mineral-rich soils in tropical regions be mined? Why or why not?

Name ______________________ Date ____________ Period ________

Activity 10A

Muffin Mining Lab

Objectives

- Determine the lasting impacts of mining and work to reclaim the mined area.
- Understand how much impact mining can have based on the materials you choose to take away.

Materials

- Blueberry or chocolate chip muffin (1 per student)
- Toothpicks
- Paper plate
- Instructions

The muffin contains an important mineral which is highly sought after by people around the world. Carefully remove as much of the mineral as possible, by whatever means is necessary. Remember that the "muffin area" will need to be returned to its original state without the mineral once you are done mining.

1. Draw a side view of your muffin before excavation begins.
2. What process(es) did you use to remove the mineral from your muffin?

   ____

3. What is the estimated percentage of mineral you removed from your muffin? Where only whole berries/chips valuable or was the surrounding residue also considered important?

   ____

   ____

4. After reclaiming the mined site, draw what your muffin looks like.

5. How does mining for a mineral below the surface impact the overall location?

   ____

   ____

6. What could you have done differently to create less of an impact on the environment?

   ____

   ____

Name ______________________ Date ______________ Period __________

Activity 10B

Soil Erosion

Objectives

- Observe the effects of water on bare soil and soil with vegetation or other cover.
- Understand how soil can be protected from erosion.

Materials

- grass sod (or an area where grass can be dug up)
- hole punch or awl
- measuring cup
- mulch, leaves, and/or other organic matter
- potting soil
- scissors
- string
- three empty small water bottles per group
- three empty two-liter bottles per group
- timer (one per group)

Step One: Prepare Soil Containers

1. Remove and discard the bottle caps from each of the two-liter bottles.
2. Cut each of the three two-liter bottles so they are open along the length of one whole side.

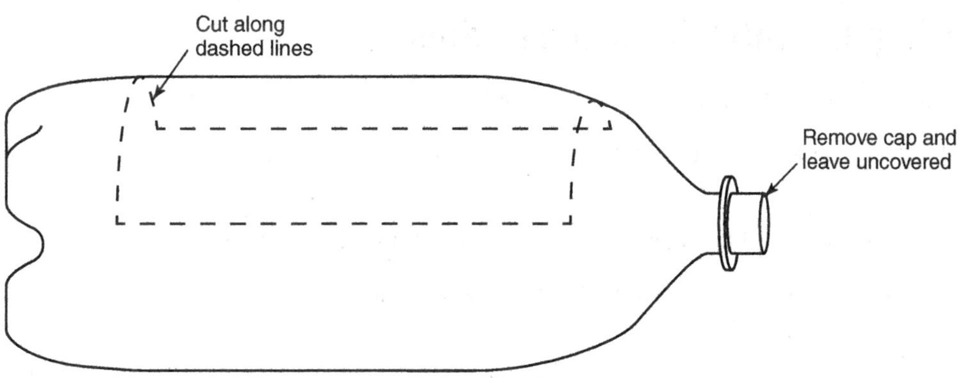

Goodheart-Willcox Publisher

Step Two: Fill the Soil Containers

1. Fill two of the bottles with potting soil so that the top of the soil is level with the opening at the top of the bottle.
2. Top off one of the bottles containing soil with mulch.
3. In the third bottle, pack the grass/sod with soil attached into the cut opening.
4. Place the three bottles along the edge of a desk or table with the open ends hanging over the edge.

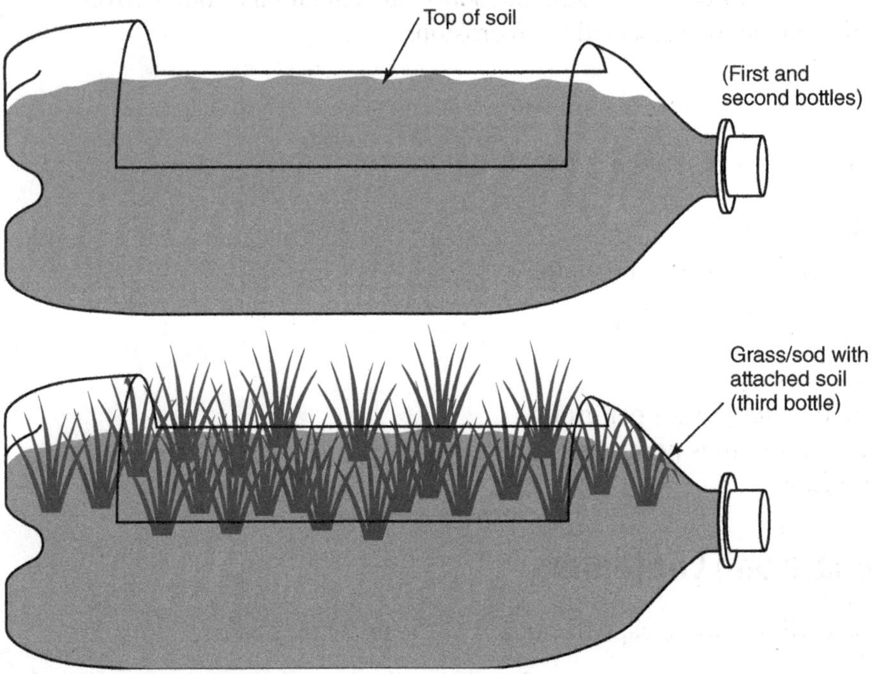

Goodheart-Willcox Publisher

Step Three: Prepare Water Collection Devices

1. Cut and discard the top half of all three water bottles.
2. Punch a hole on one side of the bottle (near the cut line). Punch another hole directly across from the first hole.
3. Cut a 10″ piece of string and tie it securely through the punched holes.
4. Hang the water collection devices from the open ends of the two-liter bottles.

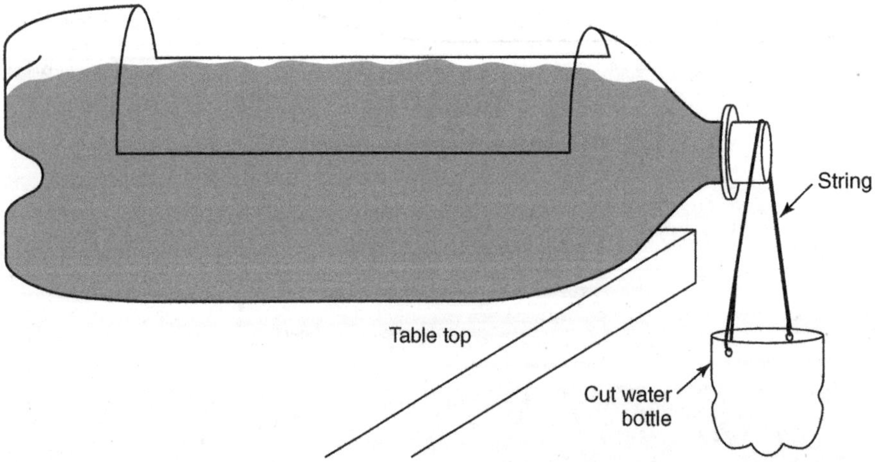

Goodheart-Willcox Publisher

Name ______________________________

Step Four: Conduct the Experiment

1. You will need to observe water flow in each bottle and record the data on the chart provided (following). Before pouring water into the containers, review the table to ensure you observe and record the necessary data.
2. Set your timer.
3. Activate the timer as you begin slowly pouring two cups of water into the first two-liter bottle (at the end opposite the opening). Record the data.
4. Repeat the process for each of the bottles.

Treatment	Time for Water to Reach Collection Device	Ammount of Water in Collection Device (use measuring cup)	Water Sediment Observations
Grass			
Mulch			
Soil only			

Step Five: Replicate the Experiment

1. Empty the water collection devices.
2. Set your timer.
3. Activate the timer as you begin slowly pouring two cups of water into the first two-liter bottle (at the end opposite the opening). Record the data.
4. Repeat the process for each of the bottles.

Treatment	Time for Water to Reach Collection Device	Ammount of Water in Collection Device (use measuring cup)	Water Sediment Observations
Grass			
Mulch			
Soil only			

Step Six: Conclusion

1. Which treatment has the most soil erosion?

2. Which treatment has the least soil erosion?

3. How does this lab relate to the methods farmers use to prevent soil erosion?

Name ______________________________ Date ______________ Period ________

CHAPTER
11
Controlling Soil Erosion

Chapter Review

Carefully study the chapter and then answer the following questions.

1. _____ What purpose does the NRCS Land Capability Classification serve for agriculturists?
   A. To determine which areas of land should be left in their natural state
   B. To determine which areas of land are best used for cultivation
   C. To determine which practices will keep their land productive
   D. All of the above.

2. _____ The plant material left in place on the field after harvest is referred to as _____.
   A. crop residue
   B. mulch residue
   C. harvest residue
   D. All of the above.

3. _____ Which of the following practices enable today's farmer to reduce the need for multiple tillage passes?
   A. Strategic crop rotations
   B. Proper weed monitoring
   C. Targeted pesticide applications
   D. All of the above.

4. _____ Which of the following are drawbacks of no-till farming?
   A. Pest control requires herbicides and pesticides
   B. Less moisture retention and more compaction
   C. More machinery and fuel required
   D. Increases the chance of soil erosion

5. _____ *True or False?* The trees or shrubs used in a shelterbelt are planted in an offset pattern to maximize its effectiveness.

For questions 6–8, match the riparian zones to the corresponding description.

A. Zone 1
B. Zone 2
C. Zone 3

6. _____ This zone is the widest and designed to slow runoff water and trap sediments, as well as absorb fertilizers in runoff water.

7. _____ The primary functions of plants in this zone are streambank stabilization, water shading, and contributing carbon to the stream.

8. _____ The primary functions of this zone are to remove contaminants from runoff water and to provide cover and habitat for wildlife.

9. _____ *True or False?* Drainage tile is installed following the contours of the soil.

10. _____ Which of the following is a benefit of tile drainage?
    A. Improved aeration and reduced soil erosion
    B. Improved productivity and crop yields
    C. Soil salinity reduction
    D. All of the above.

11. _____ What purpose does riprap serve along streambanks?
    A. To slow the flow of water
    B. To enhance the aesthetics
    C. To change the water's direction
    D. All of the above.

12. _____ *True or False?* A gabion may be filled with riprap, rock, or broken chunks of concrete and installed along a shoreline or streambank.

Vocabulary Review

Match each type of land cultivation to the most appropriate description.

A. contour tillage
B. conventional tillage
C. maximum tillage
D. mulch tillage
E. no-till
F. reduced tillage
G. ridge tillage
H. zone tillage

1. _____ The grower uses a sequence of tillage procedures to prepare the soil for planting and for weed control.
2. _____ A farmer tills the soil only in the rows where the crop will be planted.
3. _____ The practice of planting a field without plowing under the previous crop's residue.
4. _____ A cultivation system in which minimum tillage is used.
5. _____ Tillage that follows the lines of the landscape.
6. _____ Farmers create mounds of soil that run the length of the field.
7. _____ The grower tills the entire field with a tillage tool that minimally disturbs the soil.

Name ____

Match each of the buffers and structural conservation practices to the appropriate description.

A. fencerow
B. grassed waterway
C. movement corridor
D. riparian zone
E. subsurface drainage
F. surface drainage
G. terrace
H. water and sediment control basin
I. windbreak

8. _____ A broad and shallow area of grass planting in the low-lying channels of a field where water tends to flow.

9. _____ A grassed or wooded area alongside a waterway that provides habitat for wildlife and protection of the waterway from pollution and sediment.

10. _____ Channeling or removing aboveground water from a given area, such as a crop field.

11. _____ A natural area that allows wildlife to move safely from one habitat area to another.

12. _____ An earth embankment constructed across a slope to form a sediment trap and water detention basin.

13. _____ An elevated portion of the field that reduces the grade or slope of the field.

14. _____ Plantings of trees or shrubs along the edge of fields that block or reduce the force of wind flow across the field.

15. _____ Removing water from below the soil surface.

16. _____ The narrow bands of grass on either side of a fence surrounding crop fields.

Thinking Critically

1. How can the actions of an individual make a difference in the effort to help prevent soil erosion?

2. How would you present the soil erosion issue and its prevention to junior high students?

3. How does a nation that destroys its soil destroy itself?

Name ______________________ Date ____________ Period ________

Activity 11A

Find the Science

Objectives

- Understand how different science fields are integrated into conservation and preservation efforts.
- Identify the basic areas covered in different fields of science.

Materials

- worksheet
- internet access

Instructions

List three scientific fields involved for each of the resources, activities, and areas listed. Justify your answers by explaining how the science is integrated into the application.

Example: Study of invasive species: wild boar

Scientific field: *Zoology*

Justification: *The physiology of wild boar enables them to reproduce quickly.*

Scientific field: *Chemistry*

Justification: *Wild boar have a chemical digestive system that allows them a widely varied and adaptable diet.*

A. **Scientific field:** ____________________

Justification: ____

____

1. Clearing diseased trees

A. **Scientific field:** ____________________

Justification: ____

____

B. **Scientific field:** ____________________

Justification: ____

____

C. **Scientific field:** ____________________

Justification: ____

____

2. Designing and building a fish hatchery

A. **Scientific field:** ______________________

Justification: ______________________

____

B. **Scientific field:** ______________________

Justification: ______________________

____

C. **Scientific field:** ______________________

Justification: ______________________

____

3. Creating and operating a breeding program for endangered animals

A. **Scientific field:** ______________________

Justification: ______________________

____

B. **Scientific field:** ______________________

Justification: ______________________

____

C. **Scientific field:** ______________________

Justification: ______________________

____

4. Studying an insect infestation of forest trees and creating a treatment program

A. **Scientific field:** ______________________

Justification: ______________________

____

B. **Scientific field:** ______________________

Justification: ______________________

____

C. **Scientific field:** ______________________

Justification: ______________________

____

Name ______________________ Date ____________ Period ____________

Activity 11B

Designing Scientific Experiments

Objectives

- Identify the steps of the scientific method.
- Understand how to design a scientific experiment.

Materials

- worksheet

Instructions

Answer the following questions.

1. List the steps of the traditional scientific method.

2. What is the difference between an independent variable and a dependent variable?

Catfish Tolerance

Read the following scenario. Answer the questions that follow.

You are concerned about the death rate of catfish in your aquaculture operation in tanks containing a high amount of blue-green algae growth. You collect some of the algae and send it to the university to have it identified. You find that it is an algae called *Anabaena. Anabaena* is known to be toxic to fish. You design an experiment to test how much of the algae the fish can stand before they die. You obtain 6 large aquariums (all the same size) and fill them with water taken from a healthy catfish pond at your operation. You put each aquarium on the same bench in the laboratory, where the light and temperature values are identical. You let the water stand for one day before starting the experiment.

- Aquarium #1 gets 2 grams of algae and no fish
- Aquarium #2 gets 20 small catfish and no algae
- Aquarium #3 gets 20 catfish and 2 grams of algae
- Aquarium #4 gets 20 catfish and 4 grams of algae
- Aquarium #5 gets 20 catfish and 8 grams of algae
- Aquarium #6 gets 20 catfish and 16 grams of algae

The aeration rate of each aquarium is identical. You make two observations at the same time each day for two weeks. You keep track of the numbers of fish in each tank that die.

1. What is the problem?

____

____

____

____

2. What is your hypothesis?

____

____

____

____

3. What is the independent variable?

____

____

____

4. What is the dependent variable?

____

____

____

Name ____

Crop Site and Growth

Read the following scenario. Answer the questions that follow.

You have planted corn in a field with clay soil. The field is predominantly flat with a slight slope at one end where a creek borders the field. You notice that germination and growth is slowest in the flattest portion of the field. You notice good germination and rapid growth on the slight hill which meets the creek. You wonder why there are differences in growth and germination in different parts of the field. Is it due to the extra water near the creek? Is there something different about the soil in that portion of the field, or is it due to the difference in slope (north, south, east, west)?

1. What is the problem?

____

____

____

____

2. Write a hypothesis for this problem.

____

____

____

____

3. What are the constants?

____

____

____

____

4. What is the independent variable?

____

____

____

5. What is the dependent variable?

____

____

____

Scientific Methods

1. Is the scientific method just for scientists? Why is it important that a non-science student understand the scientific method? How do you think that this method can be used in everyday life?

2. Is there only one method of "doing science"? Explain your answer.

3. Does the method differ among fields of science? If so, how?

Name ______________________________ Date ______________ Period __________

Activity 11C

Communicating Results

Objective

- Understand how to graph and communicate results of a survey or other types of study.

Materials

- worksheet

Instructions

Read the following scenario. Use the information provided to create the charts and graphs requested.

Scenario

A group of high school students wanted to examine the familiarity of freshmen and sophomores with natural resources and related fields. They developed a five-question survey and randomly selected 10 students from each grade to complete the survey. The results of the survey are shown in the following chart.

Student #	Grade	GPA	On a scale from 1-10, how involved are you in natural resources?	Do you want to pursue a career in natural resources?	Are you an FFA member?	What type of setting do you live in ? (rural, suburban, or urban)
1	9	3.5	6	Yes	Yes	Suburban
2	9	3.1	4	No	No	Urban
3	9	2.0	1	No	No	Rural
4	9	4.0	10	Yes	Yes	Rural
5	9	3.0	5	Maybe	No	Suburban
6	9	1.5	10	Yes	Yes	Rural
7	9	3.4	8	Maybe	Yes	Urban
8	9	3.7	9	Yes	Yes	Urban
9	9	4.0	3	No	No	Urban
10	9	3.2	7	Maybe	No	Suburban
11	10	3.4	9	Yes	Yes	Urban
12	10	1.9	6	Maybe	No	Suburban
13	10	2.6	2	No	No	Rural
14	10	3.9	10	Yes	No	Rural

Continued

Goodheart-Willcox Publisher

Student #	Grade	GPA	On a scale from 1-10, how involved are you in natural resources?	Do you want to pursue a career in natural resources?	Are you an FFA member?	What type of setting do you live in ? (rural, suburban, or urban)
15	10	2.8	5	No	No	Urban
16	10	3.3	6	Maybe	Yes	Suburban
17	10	3.0	9	Maybe	Yes	Suburban
18	10	3.6	8	Yes	Yes	Rural
19	10	4.0	1	No	No	Urban
20	10	2.4	9	Maybe	No	Rural

Goodheart-Willcox Publisher

1. Draw a pie chart showing the breakdown of students who took this survey based on the setting in which they live.
2. Draw a bar chart showing student involvement in agriculture (use a scale of 1–10).
3. Create a scatter plot depicting the GPAs of the 20 students who completed the survey.

Name ______________________ Date ______________ Period __________

CHAPTER 12 Mining of Natural Resources

Chapter Review

Carefully study the chapter and then answer the following questions.

1. _____ *True or False?* Mining is the extraction of valuable materials from Earth.

2. _____ *True or False?* The earliest mining in the United States was performed shortly after the colonists arrived on the continent.

3. _____ Which of the following is a diverse metal with conductive and anticorrosive properties that contributes about $10 billion to the US economy?
   A. Quartz
   B. Gold
   C. Copper
   D. Iron

4. _____ Which of the following is a precious metal that contributes about $8 billion to the US economy?
   A. Gold
   B. Silver
   C. Iron
   D. Copper

5. _____ Which of the following metals is used to make steel and is the fourth most abundant element on Earth?
   A. Gold
   B. Iron
   C. Platinum
   D. Silver

6. _____ Which one of the following minerals contains silicon, is found in most sand, and is commonly used as an abrasive?
   A. Limestone
   B. Dolomite
   C. Vegamite
   D. Quartz

7. _____ Which of the following is true about crushed stone mined in the United States?
   A. It is commonly referred to as gravel.
   B. Most of it is trap rock, granite, dolomite, or limestone.
   C. It is a main component of the construction industry.
   D. All of the above.

8. _____ *True or False?* Powdered cement, which is mixed with water, sand, and crushed stone to form mortar, is derived from limestone that has been heat treated.

9. _____ Which of the following fossil fuels accounts for 15% to 17% of our nation's power production?
A. Coal
B. Natural gas
C. Crude oil
D. Gasoline

10. _____ About how much does the mining and sales of coal contribute annually to the nation's economy?
A. $10 billion
B. $22 billion
C. $30 billion
D. $63 billion

11. _____ About how much does the mining and sales of natural gas contribute annually to the nation's economy?
A. $30 billion
B. $85 billion
C. $185 billion
D. $385 billion

12. _____ Which of the following controversial mining methods is commonly used to extract natural gas?
A. Hydrodrilling
B. Hydrofracking
C. Hydroflurrying
D. All of the above.

13. _____ Which of the following results of mining causes the loss of biodiversity?
A. Overburden removal
B. Direct pollution of flora and fauna
C. Changes in water pH
D. All of the above.

14. _____ *True or False?* Sinkholes caused by mining operations occur at the site of the mining operation and in surrounding areas.

15. _____ Which of the following are the waste materials from mining operations?
A. Extractions
B. Tailings
C. Headings
D. Endings

Name ______________________________

Vocabulary Review

Match the mining techniques and terms to the appropriate definition or description.

A. fracking
B. in-situ mining
C. mountaintop removal mining
D. open-pit mining
E. placer mining
F. quarry
G. strip mining
H. subsurface mining

1. _____ A surface mining technique in which resources are removed from an open pit that is often referred to as a quarry.

2. _____ A large open pit from which minerals are mined.

3. _____ A form of surface mining in which explosives are used to remove the summit of a mountain.

4. _____ A surface mining technique in which a seam of mineral is extracted by first removing a long strip of a relatively thin layer of overburden.

5. _____ Mining minerals in a creek or streambed.

6. _____ A mining technique in which tunnels and shafts are dug into Earth to reach mineral deposits that are too deep for surface mining.

7. _____ A mining process in which an acidic brine is used to dissolve minerals to enable their extraction from the ground and rocks.

8. _____ A mining process in which a sand slurry is forced down the well with hydraulic pressure to break the rock and release the natural gas, which rises toward the surface.

Thinking Critically

1. Explain how acid rock drainage occurs.

2. Identify three mine reclamation activities and explain how they help maintain or return the area to its natural state.

3. Are diamonds and other precious stones made by people *not* considered minerals? Why or why not?

4. Is the ice in a glacier considered a mineral? Why or why not?

5. Why are some minerals harder than others?

Name ______________________ Date ____________ Period ________

Activity 12A

Minerals in Everyday Objects

Objective

- Identify objects that include various minerals in their manufacture.

Materials

- internet access
- worksheet

Instructions

Research the minerals listed and find 2–3 objects that include the minerals in their manufacture and/or their operation.

Mineral or Mineral Product	Uses
aluminum	
beryllium	
chromium	
copper	
fluorite (fluoride)	
gold	
gypsum	
halite (salt)	
iron	

Continued

Mineral or Mineral Product	Uses
lead	
mica	
molybdenum	
nickel	
perlite	
phosphate	
potassium (potash)	
silica	
silver	
sulfur	
talc	
titanium	
tungsten	
zinc	
zirconium	

Name ________________________________ Date ______________ Period __________

Activity 12B

Mineral Scavenger Hunt

Objectives

- Identify objects that include various minerals in their manufacture.
- Be the first team to find all the minerals and receive a prize.

Materials

- camera (cell phone or digital)
- internet access
- worksheet

Preparation

Divide the class into groups of 4–5 people. Write the names of your teammates in the space provided.

____

____

____

____

____

Instruction

Your instructor will provide a time limit for you to find at least one item for each mineral listed. You may add the items you listed in Lab 12A Minerals in Everyday Objects, to the following chart. Be respectful of other classes in session and do not speak loudly or run in the hallways. Do not go outside unless your instructor has given you permission.

Procedure

1. Determine who will be creating the master photo list for your group and provide each member with that person's email or cell phone number.

   Image collector:________________________

2. The person creating the master list can create a Word file ahead of time into which they will be inserting the photos. The list should have the minerals listed in the same order as the chart and the photos should be inserted next to the mineral name. This file should also include the group members' names.
3. Plan a strategy as to how you will divide the list amongst the group.
4. Email or text each picture to the person assigned to create the master photo list. Let every member of the group know you have found something that qualifies when you find it so that no one continues looking for it.
5. Email your document to your instructor after you have located and photographed all the items.
6. The first team to complete the assignment with 100% accuracy will win the prize.

Indicate the items found or add a qualifying object to the items listed in the Uses/Objects column.

Mineral or Mineral Product	Item Found (Yes/No)	Uses/Objects	Object Location
aluminum		aluminum foil, cosmetics, beverage cans, deodorant, hand lotion. antacids, cooking pots	
beryllium		fluorescent lamps	
chromium		chrome fixtures (cars, bicycles, lamps, kitchens), stainless steel	
copper		wires, pipes, cooking pots, old gutters and roofs, brass, pennies	
fluorite (fluoride)		toothpaste, drinking water	
gold		dentistry, jewelry, computers, electronics	
gypsum		wallboard, plaster	
halite (salt)		table salt, food preservatives, de-icers	
iron		cosmetics, hair dye, steel, wrought iron	
lead		car batteries, computers, fuel tanks, TV tubes, leaded glass, x-ray shields, fishing sinkers	
mica		sheetrock, paints, hair dye, cosmetics, soap, electronics, jewelry	
molybdenum		fertilizer, filament supports in lightbulbs, steel	

Goodheart-Willcox Publisher

Continued

Name ____

Indicate the items found or add a qualifying object to the items listed in the Uses/Objects column.

Mineral or Mineral Product	Item Found (Yes/No)	Uses/Objects	Object Location
nickel		nickel coins, stainless steel, alnico magnets, sheetrock	
perlite		gardening	
phosphate		fertilizer, dishwashing detergent, laundry detergent	
potassium (potash)		fertilizer, toothpaste	
silica		computer chips, glass, cosmetics, antacids, paint, laundry detergent, drain cleaner, quartz watches	
silver		photography developer, jewelry, electronics, silverware, dentistry	
sulfur		fertilizers, matches, car tires	
talc		baby powder, cosmetics, antacids, sheetrock, primer	
titanium		cosmetics, and lotion, soap, toothpaste, hair dye, bug spray, primer, paint	
tungsten		filament in lightbulbs, drill bits (tool steel)	
zinc		sunblock, fertilizer, cosmetics, dandruff shampoo, pennies, galvanized metal, brass, dry-cell batteries	
zirconium		deodorant, jewelry	

Goodheart-Willcox Publisher

Notes

Name ______________________ Date ____________ Period ________

CHAPTER 13 Water Supply

Chapter Review

Carefully study the chapter and then answer the following questions.

For questions 1–9, identify the processes and states of water in the following illustration.

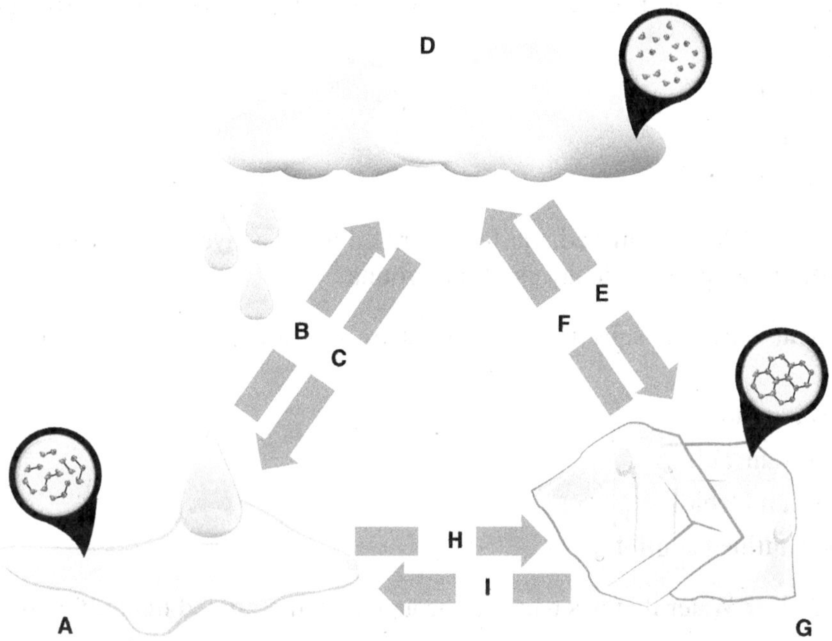

Designua/Shutterstock.com

1. _____ condensation
2. _____ deposition
3. _____ evaporation
4. _____ freezing
5. _____ gas
6. _____ liquid
7. _____ melting
8. _____ solid
9. _____ sublimation

10. _____ Why are frost heaves beneficial to water movement?
A. They cause water condensation.
B. They help water accumulate on the ground.
C. They allow water to move through the ground.
D. It is not beneficial to water movement.

11. _____ In which of the following is water located?
A. Below ground
B. In the atmosphere
C. On the ground surface
D. All of the above.

12. _____ How many gallons of freshwater used daily in the United States come from groundwater?
A. 198 million
B. 95 million
C. 82 million
D. 20 million

13. _____ *True or False?* An unconfined aquifer is formed when a layer of impermeable rock separates the water from more permeable layers of the ground.

14. _____ The point at which the subsurface materials are saturated with groundwater is referred to as _____.
A. the water table
B. an artesian well
C. the ground table
D. an unconfined aquifer

15. _____ *True or False?* Water that has trickled through soil and pooled in aquifers is groundwater.

16. _____ A natural area where groundwater flows onto the land surface is called a _____.
A. fen
B. stream
C. spring
D. well

17. _____ Which of the following often contaminate groundwater?
A. Landfills and septic systems
B. Acid rain and underground fuel tanks
C. Household products and fertilizers
D. All of the above.

18. _____ The lower end of a river where the river meets the sea is a(n) _____.
A. confluence
B. headwater
C. estuary
D. tributary

Name ______________________________

19. _____ The location where two rivers converge is a _____.
   A. confluence
   B. headwater
   C. estuary
   D. tributary

20. _____ A river or stream that flows into a larger body of water is a(n) _____.
   A. confluence
   B. headwater
   C. estuary
   D. tributary

21. _____ The beginning of a stream or river is the _____.
   A. confluence
   B. headwater
   C. estuary
   D. tributary

22. _____ *True or False?* There are fears that conflicts over water use will increase as populations increase and more water is needed for irrigation and human consumption.

23. _____ The land-based rights that refer to the allocation of water for landowners whose property borders a body of water are _____.
   A. prior appropriation water rights
   B. in-stream water rights
   C. riparian water rights
   D. individual water rights

24. _____ The water rights that entitle the first person to access the water with the right to reasonable use of the water as needed for purposes beneficial to themselves are called _____.
   A. prior appropriation water rights
   B. in-stream water rights
   C. riparian water rights
   D. individual water rights

25. _____ The water rights that only protect water in a stream and prohibit its diversion for other usage are _____.
   A. prior appropriation water rights
   B. in-stream water rights
   C. riparian water rights
   D. individual water rights

Vocabulary Review

Match the processes of the hydrologic cycle with the appropriate description.

A. accumulation
B. condensation
C. evaporation
D. infiltration
E. precipitation
F. transpiration

1. _____ The process in which liquid water is changed to vapor.
2. _____ The process in which plants absorb water through their roots and release it as vapor from their leaves.
3. _____ The process in which water vapor in the air is changed from a gas to a liquid.
4. _____ The process in which liquid water enters soil or rock.
5. _____ Water released from the atmosphere that falls in liquid or solid form.
6. _____ When water gathers in large quantities in liquid or solid form, such as in rivers.

Match the following terms to the appropriate description.

A. bog
B. fen
C. forest swamp
D. marsh
E. ocean
F. pool
G. river
H. sea
I. strait
J. stream
K. swamp
L. watershed

7. _____ A stream of water that flows into a sea, ocean, lake, or another river. It may also run dry at the end of its course.
8. _____ A forested wetland formed in a low-lying area that is constantly saturated with water.
9. _____ A length of water connecting two bodies or water, especially seas.
10. _____ A small, narrow river which is also referred to as a creek, brook, or kill.
11. _____ A wetland characterized by its spongy peat, acidic conditions.
12. _____ A wetland formed in a low-lying area that floods only at certain times of the day or year.
13. _____ A wetland that is similar to a bog but is usually less acidic and higher in nutrient concentration due to flowing water.
14. _____ An area of calm and slow moving or standing water.
15. _____ Forested wetlands formed in low-lying areas that are constantly saturated with water.
16. _____ One of the five largest saltwater bodies of water on Earth.
17. _____ Part of an ocean that is partially enclosed by land.
18. _____ A natural or manmade area of land that drains water from precipitation, surface water, and groundwater through a series of pathways to a common outlet.

Name ______________________________

Thinking Critically

1. In what ways do the people in your community rely on water?

2. Would you drink water that is cleaned wastewater? Why or why not?

3. Would you eat food that is irrigated with treated wastewater? Why or why not?

Notes

Name ______________________ Date ______________ Period __________

Activity 13A

The Water Cycle

Objective

- Understand the processes in the water cycle.

Materials

- clear plastic wrap
- dry ceramic mug
- large, clear glass bowl
- string or large rubber bands
- water

Step One

1. Draw and label the water cycle.

Step Two

1. Place the ceramic mug in the bottom of the clear glass bowl.
2. Pour water into the bowl around the mug until it is about 1/4 full. *Do not get water in the ceramic mug.*
3. Use a large rubber band or string to secure plastic wrap over the top of the bowl.
4. Place the bowl in a sunny location for several hours.
5. What type of water is simulated by the water in the bowl?

   ____

   ____

6. What is happening as the water accumulates on the plastic wrap at the top of the bowl?

   ____

   ____

   ____

Step Three

1. Leave the bowl in a sunny location. Check the bowl after 24 hours. Has any water accumulated in the ceramic mug? If so, how much?

2. Why is water accumulating in the mug?

3. What can be done to increase the amount of water accumulating in the mug?

Name ______________________ Date ______________ Period __________

Activity 13B

Taste Testing Drinking Water

Objectives

- Identify the difference in taste of water from various sources.
- Indicate whether the cost of bottled water justifies its purchase and the waste it produces.

Materials

- crackers (unsalted)
- small paper cups
- worksheet
- water from 10 sources
- 10 water pitchers (labeled 1–10)
- internet access

Instructions

In this activity, you will be comparing the taste of water from 10 sources and rating the water on a scale of 1–10, with 10 being the best. Complete the chart on the following page. Eat a cracker before you taste each type of water. After you are finished tasting the water samples, answer the following questions.

1. Which water sample did you like best? Explain why you preferred this sample.

 ____

 ____

 ____

 ____

 ____

2. What is the source of your preferred sample? Does this surprise you? Why or why not?

 ____

 ____

 ____

 ____

3. Was your preferred sample bottled water?

 ____

 If not, have you changed your view on buying bottled water, rather than using a refillable bottle? Explain your answer.

 ____

 ____

 ____

 ____

Sample Number	Type of Water	Source (location)	Types of Additives	Rating
1				
2				
3				
4				
5				
6				
7				
8				
9				
10				

For over thirty years, tap waters from all around the world have competed for the Best Municipal Water at the Berkeley Springs International Water Tasting. This judging is considered "the world's most prestigious water tasting." Judging is conducted considering taste, odor, mouthfeel, and aftertaste. Research online to see who the most recent winners have been and identify three adjectives to describe the water from that location.

1. Would any of the water samples you selected have similar adjectives to describe the water quality?

Name ______________________ Date ______________ Period __________

Activity 13C

Water Olympics

Objectives

- Observe properties of water.
- Explain how water properties affected the results of each activity.

Materials*

- activity worksheet
- cardboard
- eyedropper
- pennies
- plastic cup
- quarters
- scissors
- liquid soap
- tray of water
- paper clips
- fork
- magnifying glass

*Each team will need a complete set of materials.

Instructions

Write the names of your group members in the space provided.

____

____

____

____

Event 1: On the Water!

1. Fill a clear plastic cup with water until it is even with the rim. Add change—quarters, dimes, nickles, and then pennies, one at a time. Keep track of the total number of coins added.

   A. Quarters added: ______________ C. Nickels added: ______________

   B. Dimes added: ______________ D. Pennies added: ______________

   E. Total amount of money added: ______________________________

2. Continue adding water until water spills over the side.

3. Describe the surface of the water.

4. Which property or properties of water caused the results?

Event 2: The Balancing Act

1. Using an eyedropper, place as many drops of water on the penny as possible without spilling over the edge. Keep track of the number of drops. (Hint—the eyedropper should be straight up and down to produce an even drop each time.)
2. Continue until water spills over or the water drop collapses.
3. Describe or draw how the water appeared on the penny before the drop collapsed or spilled over.

4. Which property or properties of water caused the results?

Event 3: Bubble-Powered Boat!

1. Cut a boat shape from a piece of cardboard.
2. Cut out a small notch in the rear of the boat.
3. Place the boat in a tray of water and drop 1 drop of liquid soap into the notch. Describe what happens.

4. What caused the boat to move?

Name ______________________________

5. Try returning the boat to the start and add another drop of soap. Were the results the same?

____

____

____

6. Place a drop of water on the table and put a soap chip in the drop. What happened to it when you put the soap chip in it?

____

____

____

7. Which property or properties of water caused the reaction?

____

____

Event 4: Clipping Along!

1. How many paper clips do you think your team can suspend on the surface of water?

   No. of clips: ______________________

2. Lay the paper clip on the prongs of the fork and set the clip on the water surface.
3. Use the magnifying glass to observe the surface of the water where it comes in contact with the paper clip.
4. Draw a picture or describe what this looks like.

____

____

____

____

5. How many paper clips did you suspend on the water surface?

   No. of clips: ______________________

6. Which property or properties of water allowed you to suspend the paper clips?

____

____

____

Notes

Name ______________________ Date ____________ Period ________

CHAPTER 14 Water Quality

Chapter Review

Carefully study the chapter and then answer the following questions.

1. _____ The basic unit of an element is a(n) _____.
   A. neutron
   B. molecule
   C. atom
   D. electron

2. _____ A group of atoms bonded together is a(n) _____.
   A. electron
   B. molecule
   C. proton
   D. atom

3. _____ *True or False?* The oxygen atom and the hydrogen atoms in water (H_2O) share their electrons.

4. _____ *True or False?* The negative and positive charges of subatomic particles cause the particles to attract or repel each other.

5. _____ *True or False?* Water molecules are dipolar molecules because they have two hydrogen atoms for each oxygen atom.

6. _____ The bonds created between water molecules are _____.
   A. relatively weak
   B. hydrogen bonds
   C. due to electronegativity
   D. All of the above.

7. _____ Surface tension _____.
   A. enables water to flow freely
   B. makes water heavier
   C. exists when surface molecules repel
   D. All of the above.

8. _____ *True or False?* The more water moves through aeration or changes in elevation, the less oxygen it can retain.

9. _____ Which of the following retain more oxygen?

A. Saltwater and cold water

B. Warm water and freshwater

C. Cold water and freshwater

D. Saltwater and warm water

10. _____ The term pH refers to the amount of _____.

A. oxygen ions in a substance

B. hydrogen ions in a substance

C. gaseous oxygen in a substance

D. All of the above.

11. _____ Zero on the pH scale is the strongest _____.

A. acid

B. alkaline

C. base

D. alcohol

12. _____ Fourteen on the pH scale has the highest _____.

A. acidity

B. alkalinity

C. neutrality

D. concentration of hydrogen ions

13. _____ What is the pH of pure water?

A. 3

B. 7

C. 10

D. 14

14. _____ *True or False?* Most living things require a specific pH range for survival.

15. _____ Why is normal, clean rainwater slightly acidic?

A. Absorption of excess oxygen

B. Absorption of excess hydrogen

C. Absorption of CO_2 and other gases

D. All of the above.

16. _____ Rainwater with a low pH is _____.

A. acid rain

B. normal rain

C. neutral rain

D. All of the above.

Name ____

17. _____ The percentage of all the water on Earth that is fresh water is _____.

A. 0.5% to 1.0%

B. 2.5% to 3%

C. 10.5% to 12%

D. 20% to 22%

18. _____ *True or False?* Domestic water use in the United States accounts for more than 30% of our daily water use.

19. _____ *True or False?* The average US citizen uses 80 to 100 gallons of water per day for domestic purposes.

20. _____ *True or False?* The average citizen in Mali uses an average of 2.8 gallons of water per day.

21. _____ The EPA determines water quality by testing for _____.

A. disinfectants and chemicals

B. radionuclides

C. microorganisms

D. All of the above.

22. _____ *True or False?* The majority of US citizens receive water from private wells.

23. _____ *True or False?* Worldwide, agricultural use accounts for 20% of global water consumption.

24. _____ In economically strong countries with much production agriculture, the consumption of water used for irrigation _____.

A. is as high as 82%

B. may be as low as 32%

C. is about 50%

D. averages about 45%

25. _____ Agricultural water supply is drawn from _____.

A. municipal systems and some treated wastewaters

B. ponds, lakes, reservoirs, rivers, and streams

C. groundwater, irrigation ditches, and open canals

D. All of the above.

26. _____ *True or False?* In some countries, industrial water use may exceed 50% of their water consumption.

27. _____ Industrial water supply systems include _____.

A. flowing waterways

B. groundwater wells

C. bodies of surface water

D. All of the above.

28. _____ Point source pollution is from _____.

A. a single identifiable source

B. an unidentifiable source

C. many unidentifiable sources

D. All of the above.

29. _____ Water pollution changes the physical, chemical, or biological condition of the water and may be caused by _____.

A. chemicals and bacteria

B. runoff from crop fields

C. bags, bottles, and other refuse

D. All of the above.

30. _____ *True or False?* Nonpoint source pollution is easily identified, as it comes from a specific point of origin.

31. _____ Living organisms that provide evidence of the health of an ecosystem are _____.

A. small rodents

B. shoreline plants

C. bioindicators

D. All of the above.

32. _____ *True or False?* Eutrophication reduces the amount of light aquatic plants need and eventually reduces the dissolved oxygen level.

33. _____ BMIs are wildlife without backbones and _____.

A. are extremely tolerant of water contaminants

B. live in the benthic level of a water body

C. cannot be seen with the naked eye

D. All of the above.

34. _____ *True or False?* Amphibians are susceptible to mutations due to contamination because they have sensitive skin and go through metamorphosis in the outside environment.

35. _____ The most common sources of wastewater include _____.

A. domestic gray water and sewage water

B. commercial water used to clean/cool products

C. industrial water used to clean/cool products

D. All of the above.

36. _____ *True or False?* Bioaccumulation of microplastics impacts the cardia system of animals over time.

Name ____

Vocabulary Review

Match the following terms with the appropriate definition or description.

A. acid rain
B. dipolar molecule
C. dissolved oxygen (DO)
D. dredging
E. effluent
F. eutrophication
G. hydroseeding
H. mutation
I. nonpoint source pollution
J. PBTs
K. radionuclide
L. universal solvent

1. _____ Pollutants, such as pesticides, arsenic, and methylmercury, that sink and accumulate in the bed or base sediment of a water body.
2. _____ A molecule with different electrical charges on each end.
3. _____ A planting process in which a slurry of seed and mulch are sprayed over prepared ground to provide groundcover.
4. _____ The measureable amount of gaseous oxygen in liquid water.
5. _____ A process by which a body of water becomes enriched with excess nutrients that cause the rapid growth of aquatic plant life.
6. _____ Suspended solids in wastewater.
7. _____ A spontaneous change in the biological structure of an organism.
8. _____ A substance, such as water, that is capable of dissolving more substances than any other solvent.
9. _____ Rainwater with a low pH that is typically caused by atmospheric pollution.
10. _____ The removal of eroded soils or contaminated sediment from the bottom of the water body.
11. _____ An unstable atom that emits radioactivity as it attempts to become more stable.

Thinking Critically

1. Explain how and why bioaccumulation occurs.

____

____

2. Explain the main difference between the following illustrated wastewater systems. Why is this difference significant?

____

____

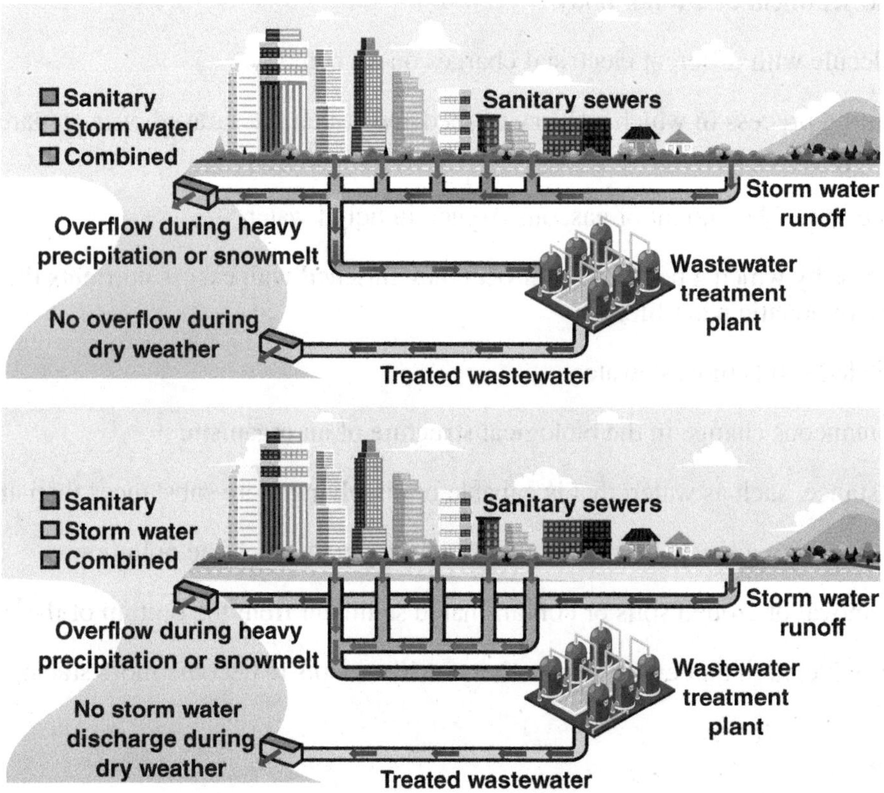

(city) Sabelskaya/Shutterstock.com; (water treatment) Macrovector/Shutterstock.com

3. How do automobiles affect storm runoff? Would an increase in the number of vehicles increase the effect? Explain your answer.

____

____

4. Describe the impact of impermeable surfaces, such as parking lots and sidewalks, on urban runoff.

____

____

5. How does US water quality compare to the water in other countries? Which countries have tap water that is considered safe for drinking?

____

____

Name ______________________________ Date ______________ Period ________

Activity 14A

Testing Water Quality

Materials

- clear jars with lids
- labels
- magnifying glass
- microscope
- notebook/sketchbook
- pond identification guide and the macroinvertebrate field guide
- stereoscope
- water meter/water test kit
- worksheet

Instructions

In this activity, you or your teacher will collect water samples from two ponds in your community. Label each sample with its location of origin. You will be observing, collecting, and recording data on the pond samples. A pond identification guide and the macroinvertebrate field guide can help to identify organisms living in the water.

Procedure

1. Look through the outside of the jar and describe the condition of the water.
   A. Are there any live organisms you can see?______________________________
   B. Does the water have an odor?______________________________
   C. Record this data in the Water Data chart.
2. Gather data on the samples, including temperature, pH, and the presence of nitrates. Record this data in your chart.
3. Use the equipment provided to identify as many organisms as you can in the two samples. This includes plants, animals, and microorganisms.
4. Sketch the organisms you see to include in your final report. Include sketches of organisms you cannot identify.
5. Briefly summarize your findings.

____

____

____

____

____

____

____

6. Use the data from your chart to write a traditional lab report and then create an infographic to share with your class. The report and infographic must contain all of your data. Include a list of reference sources.

Water Data		
Data	**Sample 1**	**Sample 2**
Location		
Initial observation		
Odor		
Organisms		
Other observations		
Temperature		
pH		
Nitrates		
Identified organisms (names)		
Other		

Goodheart-Willcox Publisher

Identified Organisms

Please label your sketches.

Unidentified Organisms

Please label your sketches.

Name ______________________ Date ____________ Period ________

CHAPTER 15 Wetlands

Chapter Review

Carefully study the chapter and then answer the following questions.

1. _____ Which of the following statements is true about wetlands?
   - A. They are vital to our ecology
   - B. They are transitional areas
   - C. They may be dry for extended periods
   - D. All of the above.

2. _____ What percentage of the animals on the endangered species list rely wholly or partly on wetlands for habitat and food?
   - A. About 25%
   - B. Nearly 50%
   - C. Nearly 73%
   - D. About 85%

3. _____ Which of the following is a role played by wetlands?
   - A. Carbon sink
   - B. Floodwater control
   - C. Food source
   - D. All of the above.

4. _____ *True or False?* The water storage and filtration capabilities of wetlands may make the quality of water flowing out of the wetland higher than the quality of the water flowing into the wetland.

5. _____ *True or False?* Due to the anaerobic conditions of wetland soils, many of the minerals are transformed into forms most usable by plants.

6. _____ *True or False?* The color and texture of sand on a beach is determined by the types of rocks found in the area and the types of minerals the rock/sand contains.

7. _____ Which of the following determines whether the soil in a wetland is mineral soil or organic soil?
   - A. The number of days it is saturated each year
   - B. The percentage (by weight) of organic materials in the O-horizon
   - C. Whether it is naturally or artificially drained
   - D. All of the above.

8. _____ Partially decomposed organic matter saturated with water is referred to as _____.
   - A. muck
   - B. top soil
   - C. peat
   - D. All of the above.

9. _____ Soils in which only one-third of the material is highly decomposed are _____.
A. saprist soils
B. hemist soils
C. fibrist soils
D. serotinous soils

10. _____ Soils that spend at least half of each year above water, and have a lower density of decaying plant matter, are _____.
A. saprist soils
B. hemist soils
C. fibrist soils
D. serotinous soils

11. _____ Soils in which two-thirds of the soil material is highly decomposed are _____.
A. saprist soils
B. hemist soils
C. fibrist soils
D. serotinous soils

12. _____ *True or False?* When cultivating sod, a layer of dried muck (soil) is also removed and the field gets lower and wetter with each harvest.

13. _____ A wetland with highly saline water that is found on coastlines is a _____.
A. groundwater wetland
B. traditional wetland
C. poor fen
D. marine wetland

14. _____ *True or False?* Coastal wetlands are areas that retain water and drain into a larger body of water, traditionally a sea or an ocean.

15. _____ A complex, underwater ecosystem built by generations of calcium carbonate secretions is a _____.
A. saltwater reef
B. coral marsh
C. coral reef
D. All of the above.

16. _____ What percentage of the world's coral reefs are in Southern Asia and the Pacific Ocean?
A. 15%
B. 32%
C. 60%
D. 70%

17. _____ Which of the following statements accurately describes characteristics of a saltwater marsh?
A. Provides habitat for many species of wildlife
B. Protects against inland flooding and soil erosion
C. Provides nursery grounds for many marine fish
D. All of the above.

Name ______________________________

18. _____ A tropical tidal wetland in which a variety of halophyte trees and shrubs thrive is a _____.
A. mangrove swamp
B. saltwater marsh
C. coral marsh
D. All of the above.

19. _____ A freshwater wetland that receives almost all of its moisture from precipitation and is comprised of moist, spongy ground and highly acidic conditions is a _____.
A. marsh
B. bog
C. swamp
D. rich fen

20. _____ A freshwater wetland that is fed mainly through groundwater and other moving water sources, such as a spring or stream, is a _____.
A. marsh
B. bog
C. fen
D. swamp

21. _____ A fen with a low pH, few nutrients, and few plants and animals is a(n) _____.
A. inland fen
B. poor fen
C. rich fen
D. freshwater fen

22. _____ A fen with a pH above 6.9, higher nutrient levels, and a greater variety of plant and animal species than *a poor fen* is a(n) _____.
A. freshwater fen
B. inland fen
C. sea fen
D. rich fen

23. _____ Freshwater marshes are typically treeless and support a variety of _____.
A. hydrophilic plants
B. herbaceous plants
C. serotinous plants
D. carnivorous plants

24. _____ A nontidal wetland that lacks flowing water and contains ocean-derived salts in small concentrations is a _____.
A. prairie pothole
B. poor fen
C. pocosin
D. prairie bog

25. _____ A short-lived or permanent wetland that consists of a hole or depression in low-lying areas where water collects and pools is a _____.

A. prairie pothole

B. poor fen

C. pocosin

D. wet meadow

26. _____ A forested wetland that relies on other freshwater sources, such as rivers or lakes, for its moisture is a _____.

A. pocosin

B. river fen

C. swamp

D. vernal swamp

27. _____ Which of the following help educate and/or raise funds for the management and restoration of wetlands?

A. Ducks Unlimited

B. Migratory Bird Hunting Stamp Act

C. Migratory Bird Act

D. All of the above.

28. _____ *True or False?* Aquatic buffers, riparian zones, and riparian buffers are undeveloped areas next to bodies of water or wetlands that provide habitat for wildlife and reduce soil erosion.

29. _____ The creation of a wetland site where one does *not* exist to replace one that is being eliminated is referred to as wetland _____.

A. enhancement

B. replication

C. creation

D. restoration

30. _____ *True or False?* Enhancements are changes made to repair or modify an existing wetland.

31. _____ When an existing wetland is converted to a different type of wetland, it is _____.

A. replicated

B. enhanced

C. reallocated

D. restored

Name ____

Vocabulary Review

Match the following terms with the appropriate definition or description.

A. anaerobic
B. deepwater habitat
C. ephemeral pool
D. freshwater wetland
E. halophyte
F. histosol
G. hydric soil
H. hydrophilic
I. mangrove
J. muck
K. ombrotrophic
L. peat
M. playa lake
N. pyrophytic ecosystem
O. riparian zone
P. serotinous
Q. topography
R. wet meadow

1. _____ A marshlike pond that receives water primarily from surface runoff and precipitation. It may periodically remain dry for up to several years.
2. _____ Organic wetland soil that is commonly referred to as peat or muck.
3. _____ A nontidal wetland made of grassland with soil that is saturated or waterlogged after extensive precipitation.
4. _____ The absence of oxygen.
5. _____ A plant that has adapted to areas with low oxygen and high concentrations of salt in the water and/or soil.
6. _____ Having an affinity for water, such as water-loving plants.
7. _____ A small, shallow depression in forests or grasslands that is usually wet only in the early spring or during the winter season.
8. _____ A soil that is saturated for long enough periods to develop anaerobic conditions.
9. _____ A strip of protected land located along water bodies and waterways that protects the water quality by preventing contamination from erosion and runoff.
10. _____ The surface configuration of an area of land.
11. _____ A type of halophyte tree or shrub that grows in mangrove swamps.
12. _____ When a wetland receives almost all of its water from precipitation.
13. _____ A wetland found inland along lakes and ponds, on the floodplains of rivers and streams, and in isolated, low-lying areas surrounded by dry land.
14. _____ An ecological adaptation in which a seed is released from a plant only after an environmental trigger, such as fire.
15. _____ An ecosystem in which many of the plants have become fire-adapted and may require fire to seed or bloom.
16. _____ An organic wetland soil created by partially decomposed organic matter saturated with water.
17. _____ An organic wetland soil that closely resembles true topsoil and is formed with decomposed vegetation.

Thinking Critically

1. Draw or obtain a map of a freshwater wetland and its surrounding area. Is the wetland near development? Identify possible sources of contaminants and other possible threats to the area.

2. Is it more difficult to restore some types of wetlands in comparison with other wetlands? Why or why not?

3. Does restoration, creation, or enhancement of a wetland compensate for loss of functions and values of another wetland? Explain why or why not.

4. Reach out to a local wetland restoration group in your community or state and learn about the projects that are underway. How will those projects impact you, your community, and the surrounding area positively and/or negatively?

Name ______________________ Date ____________ Period ________

Activity 15A

Groundwater Infiltration Lab

Objectives

- Understand how groundwater is filtered through layers of soil and stone.
- Determine which materials work well as water filters.

Materials

- "contaminated" water
- charcoal
- coffee filters
- cotton
- empty water bottles with bottoms removed (2 per student/group)
- gravel
- rubber bands
- scissors
- screen
- soil

Section 1

1. Define *infiltration*.

____

____

____

____

2. List three factors that contribute to the success of infiltration.

____

____

____

____

____

____

____

____

Section 2

Using the "bottle" diagrams, draw and label the layers you will be using in your filtration system. Each group is allowed two design attempts.

System #1

1. What do you think is the most successful part of this design?

2. Layer the materials in one of your water bottles in the same configuration as your design.
3. What was left in the water after the first filtration?

4. Rank the color on a scale of 1–5 (1 being very clear) after the first filtration.

5. What was left in the water after the second filtration?

6. Rank the color on a scale of 1–5 (1 being very clear) after the second filtration.

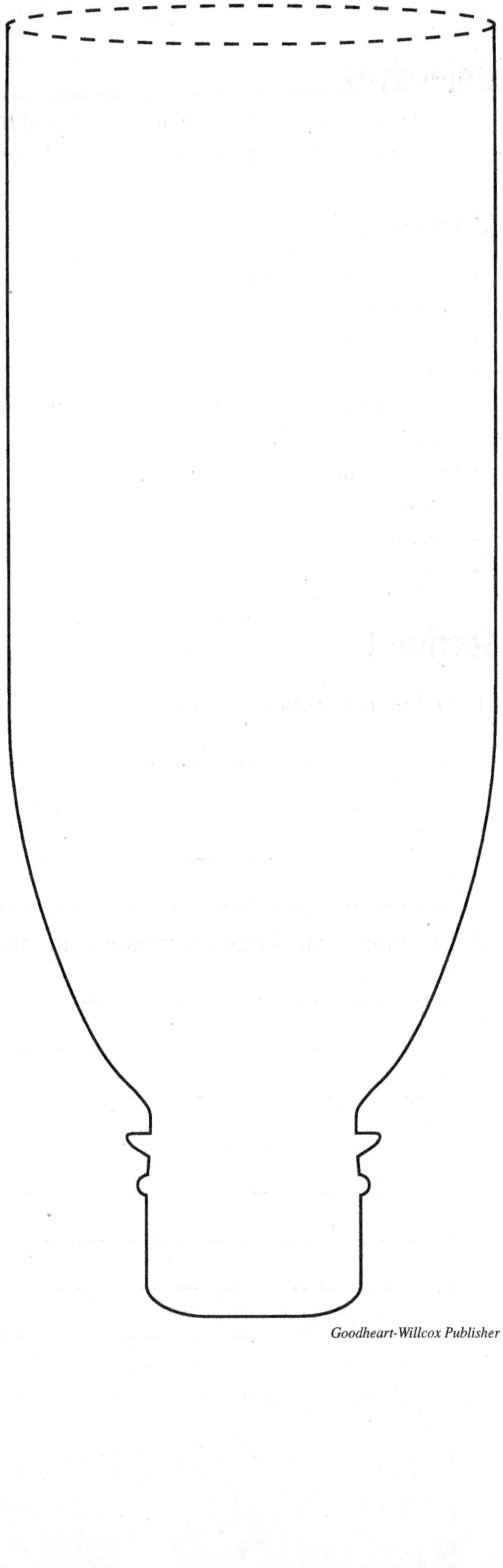

Goodheart-Willcox Publisher

Name ______________________________

System #2

1. What do you think is the most successful part of this design?

____
____
____
____
____

2. Layer the materials in one of your water bottles in the same configuration as your design.
3. What was left in the water after the first filtration?

____
____
____
____
____

4. Rank the color on a scale of 1–5 (1 being very clear) after the first filtration.

____
____

5. What was left in the water after the second filtration?

____
____
____
____
____

6. Rank the color on a scale of 1–5 (1 being very clear) after the second filtration.

____
____

Goodheart-Willcox Publisher

Section 3

1. What is the best component of filtration?

2. What was the hardest contaminant to remove?

3. Conduct an internet search to determine how activated charcoal works. Explain why activated charcoal is common in water filters.

Name ______________________ Date ____________ Period ________

CHAPTER 16 Weather and Climate

Chapter Review

Carefully study the chapter and then answer the following questions.

1. _____ *True or False?* The state of the atmosphere at a particular place with respect to meteorological conditions is an area's climate.

2. _____ *True or False?* The long-term, average weather patterns in an area over an extended period is the area's weather.

3. _____ The layer of the atmosphere that has very thin air, is greatly affected by solar activity, and is considered by some to be a part of deep space is the _____.
   A. exosphere
   B. Karman line
   C. thermosphere
   D. mesopause

4. _____ The first layer of the atmosphere that is the first layer of defense against the sun's radiation is the _____.
   A. mesosphere
   B. thermosphere
   C. stratopause
   D. exosphere

5. _____ The layer of the atmosphere that is about 22 miles (35 km) above Earth's surface and composed of a mass of protective gases is the _____.
   A. stratosphere
   B. stratopause
   C. exopause
   D. mesosphere

6. _____ The layer that begins at the stratopause, is about 32 miles (52 km) above Earth's surface, and does not support flight, is the _____.
   A. tropopause
   B. mesosphere
   C. mesopause
   D. troposphere

7. _____ The layer of the atmosphere that varies in height, begins at Earth's surface, and is the densest is the _____.

A. mesosphere
B. exosphere
C. stratosphere
D. troposphere

8. _____ *True or False?* The temperature and air pressure in the troposphere varies greatly depending on location.

9. _____ *True or False?* The ozone layer provides the least amount of protection from certain types of ultraviolet radiation.

10. _____ *True or False?* Climate change is occurring primarily due to anthropogenic actions and changes to Earth.

11. _____ The air pressure in the atmosphere varies in part because _____.

A. cold air falls
B. warm air rises
C. Earth is not equally heated by the sun
D. All of the above.

12. _____ *True or False?* In a high-pressure system, the air is moving in a counterclockwise direction and cooling as it rises.

13. _____ *True or False?* In a low-pressure system, the cooler air sinks and fans out as it reaches the ground.

For questions 14–21, identify the actions and parts of the cold front.

14. _____ altocumulus
15. _____ cirrostratus
16. _____ cirrus
17. _____ cold air
18. _____ cold front
19. _____ cumulonimbus
20. _____ warm air
21. _____ wind

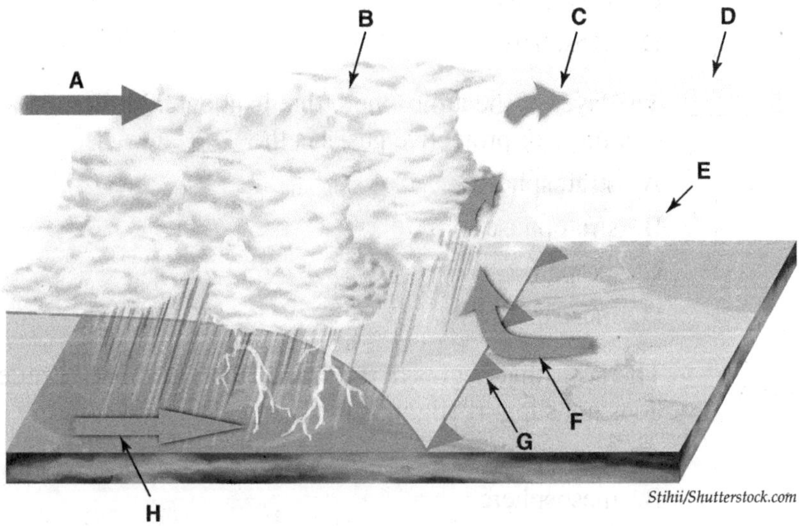

Stihii/Shutterstock.com

Name ______________________________

For questions 22–30, identify the actions and parts of the warm front.

22. _____ altostratus

23. _____ cirrostratus

24. _____ cirrus

25. _____ cold air

26. _____ nimbostratus

27. _____ stratocumulus

28. _____ warm air

29. _____ warm front

30. _____ wind

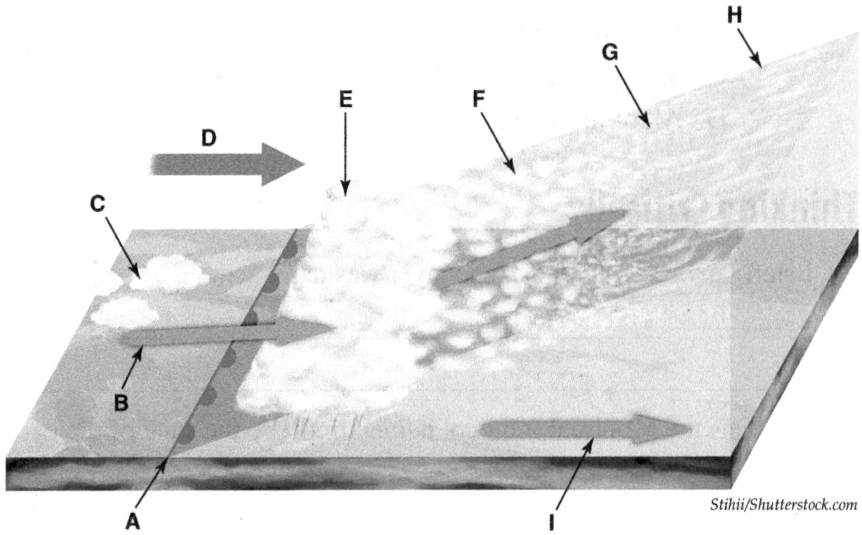

Stihii/Shutterstock.com

31. _____ Which of the following statements about trade winds is *not* true?
    A. One blows southwesterly toward the equator from the northeast in the northern hemisphere
    B. One blows northwesterly toward the equator from the southeast in the southern hemisphere
    C. They form as Earth's rotation deflects air to slant toward the equator
    D. They flow in relatively straight paths, north to south and south to north

32. _____ *True or False?* Earth's rotation causes the jet streams to flow west to east.

33. _____ Which of the following occurs when we are experiencing the weather pattern referred to as El Niño?
    A. Southern portion of the United States experiences heavy rainfall and flooding
    B. Trade winds weaken and warm water moves east toward the western coast of South America
    C. Warmer water causes a southward shift of the jet stream
    D. All of the above.

34. _____ Which of the following occurs when we are experiencing the weather pattern referred to as La Niña?
    A. The trade winds blow warm surface waters from our west coast toward Asia
    B. Trade winds weaken and warm water moves east toward the western coast of South America
    C. Trade winds are often stronger than normal and blow the warm water toward the western Pacific
    D. All of the above.

35. _____ *True or False?* The difference between a normal weather year and an El Niño or La Niña year is based on ocean temperatures.

36. _____ A prolonged period of abnormally low rainfall and the water shortage that follows is referred to as a _____.
    A. dry front
    B. drought
    C. dry habitat
    D. All of the above.

37. _____ Which of the following often occur during flooding?

A. Contaminated water supplies
B. Destroyed crops
C. Soil erosion
D. All of the above.

Thinking Critically

1. Briefly explain why the ozone layer was damaged.

____

____

2. How has the Montreal Protocol helped improve the state of the ozone layer?

____

____

3. The greenhouse effect is a natural occurrence that prevents Earth from becoming a cold planet. If it is a necessary occurrence, why is it harmful to Earth?

____

____

4. Why do winds affect weather?

____

____

5. On what basis is the intensity of tornados or hurricanes rated? How is their strength measured?

____

____

6. What is the strongest evidence that Earth's atmosphere is warming because of human activities? On what basis do those who disagree base their doubt?

____

____

7. Why do weather forecasts from different sources vary? Are weather forecasters interpreting the data differently? If so, how?

____

____

8. July 4, 2023 has been recorded as one of the hottest days in history but it probably won't hold the record for long. Why do you think this is so? How do worldwide increased temperatures impact our way of life?

____

____

Name ______________________ Date ____________ Period ________

Activity 16A

Weather Forecast

Objectives

- Compose a weather forecast for a chosen area.
- Understand which aspects of an area affect its weather.
- Write and present a weather forecast.

Materials

- activity sheet
- internet access

Instructions

Your teacher will divide the class into small groups of 3–4 people, and each group will select a city (other than their own) and determine how they will present their forecast. Answer the following questions to compile information needed to create a fictional five-day weather forecast for a chosen city.

Section One: Background Information

In order to create your forecast, you will need to identify certain aspects of the city's location, including the terrain (city proper and surrounding), proximity to larger bodies of water, latitude and longitude, and the altitude.

1. Identify your chosen city and the country in which it resides.

____

2. What is the latitude and longitude of your chosen city?

____

A. How does the latitudinal and longitudinal location of your city affect its climate?

____

____

____

____

3. What is the altitude of your city?

____

____

A. How does the altitude affect your city's climate?

____

____

____

____

4. What type of terrain does your city have? What type of terrain surrounds the city?

____

____

____

   A. How does the terrain affect your city's climate?

____

____

____

5. Are there any major bodies of water near or bordering your city?

____

____

   A. If so, how does the proximity of these bodies of water affect your city's climate?

____

____

____

6. Are there other significant factors that influence the local climate (i.e., hot springs, excessive wind, etc.)?

____

____

Section Two: Forecast Information

Using the information you gathered, create a fictional five-day forecast for your city.

1. What are the predicted average high and low temperatures for today in this location?

   High: ____________________

   Low: ____________________

2. What is the weather forecast for the next five days? List the day of the week and the corresponding forecast.

____

____

____

3. Explain how warm air and cold air fronts could influence the upcoming weather.

____

____

____

4. Will there be any precipitation? If so, how much and when?

____

____

____

Name ______________________ Date ____________ Period ________

Activity 16B

Making Lightning

Objectives

- Observe lightning formation.
- Understand actions of electrons and protons in lightning formation.

Materials

- aluminum pie pan
- foam plate
- pencil with a new eraser
- rubber gloves
- small piece of wool fabric
- thumbtack

Procedure

1. Push the thumbtack through the center of the aluminum pie pan from the bottom.
2. Push the eraser end of the pencil into the thumbtack.
3. Put the foam plate upside-down on a table.
4. Put a rubber glove on one hand. Using the gloved hand, quickly rub the foam plate with the wool for a couple of minutes.
5. Pick up the aluminum pie pan using the pencil as a handle and place it on top of the upside-down foam plate that you were just rubbing with the wool.
6. Touch the aluminum pie pan with your bare finger. What happened? Explain why this happened.

 ____

 ____

 ____

 ____

 ____

7. Touch the pie pan again. What happened? Is it what you expected?

 ____

 ____

 ____

 ____

 ____

8. Try the experiment in a darkened room and observe the results.

What Is Happening?

The "lightning" you see is caused by the movement of electrons. Electrons carry a negative electrical charge, which causes them to be attracted to protons and repelled from other electrons. Lightning occurs when the electrons in the atmosphere are attracted to the protons on the ground, or when there is a big charge difference between the clouds and Earth. When you rubbed the wool on the plate, you either put the electrons on the plate, or stripped electrons off the plate. This gives the plate a net charge, meaning it has either more electrons or fewer electrons than protons. The accumulation of electric charges has to be great enough to overcome the insulating properties of air. When this occurs, a stream of electrons (negative charges) pours down toward a high point, such as a tree, where protons (positive charges) have clustered due to the pull of the thunderhead. A bolt of lightning heats the air in its path, which causes the air to expand rapidly. Thunder is the sound caused by rapidly expanding air.

Name ______________________ Date ____________ Period ________

Activity 16C

Lightning Lab Variations

Version #1

In this activity, you will create lightning in a jar. You will create an anode and a cathode to conduct the electricity. The anode is the positively charged electrode at the bottom of the jar and the cathode is the negatively charged electrode at the top of the jar.

Materials

- aluminum foil
- dryer sheet
- glass jar with metal lid
- rubber balloon
- thumbtacks

Procedure

1. Fold a 12″ × 12″ piece of aluminum foil twice over into a smaller square. The square should fit in the bottom of the jar. It does not need to fit perfectly flat so the edges may stick up.
2. Poke several thumbtacks through the dryer sheet.
3. Set the dryer sheet over the opening of the jar with the thumbtacks pointing down inside the jar. Tighten the lid over the dryer sheet.
4. Charge the balloon by rubbing it on your hair.
5. Touch the balloon to the top of the jar and observe the reaction in the jar. What occurred? Explain the reaction.

   ____

   ____

   ____

   ____

   ____

6. If the balloon has sufficient charge, you will see flashes of lightning.

Version #2

You will use a few different materials in this version of the original lightning experiment.

Materials

- foam plate or rubber balloon
- head of hair or wool
- plastic fork or small spatula
- rubber glove
- tin foil
- wood or plastic cutting board
- cool, low-humidity day (< 45% humidity, < 75°F)

Procedure

1. Fold the aluminum foil around the plastic fork or spatula. Make it as flat as possible with rounded corners.
2. Put on one rubber glove and use your gloved hand to rub the foam plate or balloon on your hair or with the piece of wool.
3. Set the plate or balloon on the cutting board, and use your gloved hand to pick up the fork wrapped in aluminum foil.
4. Place the foil-covered part of the fork on the plate or balloon.
5. Touch the foil with your bare hand. What happened? Can you explain why this occurred?

   ____

   ____

   ____

6. Lift the fork and touch the plate again. What happened? Is it what you expected?

   ____

   ____

   ____

7. Try the experiment in a darkened room and observe the results.

Version #3

In this lightning experiment, you will create lightning in your mouth.

Materials

- dark room
- mirror
- wintergreen or peppermint Life Savers®

Procedure

1. Let your eyes adjust to the darkness.
2. Put a mint in your mouth and, while keeping your mouth open, break the Life Saver® with your teeth.
3. Did you see any sparks? If not, try it again. As you break the Life Saver®, you are breaking sugars in the candy. The sugars release electrical charges and attract the oppositely charged nitrogen in the air. When they meet, the result is the tiny sparks.

Name ________________________________ Date ______________ Period __________

Activity 16D

Tornado in a Bottle

Objectives

- Understand how a tornado forms.
- Identify the forces in play in a vortex, or why the water is spinning in the bottle.

Materials

- 2 empty and clean 2-liter clear plastic pop bottles*
- dish washing liquid (clear)
- funnel
- 3–4 tablespoons of larger glitter (do not use fine glitter)
- plastic houses from a board game (optional)
- vortex connector, or a metal washer and duct tape
- water
- water pitcher

*This experiment may also be performed with one bottle or jar.

What Is a Tornado?

A tornado is a violent atmospheric storm that forms quickly when different temperatures and humidity meet, such as when a cold front moving from the north hits warm humidity traveling from the south. This typically occurs during thunderstorms. A tornado, which is a rotating column of air called a vortex, may form when the two fronts connect and create instability in the atmosphere. A vortex is a type of motion that causes liquids and gases to travel in spirals around a center line. A change in wind direction and an increase in wind speed with increasing height creates an invisible, horizontal spinning effect in the lower atmosphere. The rising air tilts the rotating air from horizontal to vertical and the area of rotation expands 2–6 miles wide, extending through much of the storm.

Powerful and destructive tornadoes form quickly with little warning and leave large paths of destruction in minutes. Winds can reach well over 200 mph and the swath of the tornado can be wider than two miles. There is a distinct tornado season in spring and summer and, in the United States, the storms occur most frequently in tornado alley, which is in the central area of the United States. However, tornadoes can happen at any time of the year and do occur on every continent except Antarctica.

Preparation

1. Remove the labels and wash the plastic bottles.
2. Remove the bottle tops and the plastic ring that is left on the neck of the bottle.
3. Fill one of the bottles three-quarters full. If you add more water, it will be more difficult to create the vortex, and if you add less water, it will shorten its duration.
4. Add a few drops of dish washing liquid and use the funnel to add the glitter. The dish soap lowers the water's surface tension and the glitter makes the vortex easier to see. The glitter also simulates the dust and debris that is visible in a tornado. If desired, you may add the small plastic houses or other objects with the glitter.

5. Attach the vortex connector securely to the full bottle first and then attach the second (empty) bottle. The plastic vortex connector holds the bottles together and narrows the water flow as it transfers from bottle to bottle. The connector can be purchased online, at craft stores, and at teacher supply stores. A metal washer can also be used to narrow the water flow, but the bottles will have to be taped together with duct tape.
6. If you are using a metal washer, place the washer on top of the full bottle and turn the empty bottle upside-down and align the two bottles. Have someone hold the second bottle as you secure the bottles with duct tape.

What Will Happen?

1. Predict what will happen when you turn the bottles so the water-filled bottle is on top and you rotate them to create the vortex. Why does this occur?

   ____

   ____

   ____

   ____

2. Predict what will happen if you do not rotate the bottles after you turn them and the water-filled bottle is on top. Why does this occur?

   ____

   ____

   ____

   ____

Procedure

1. Check for leaks before you begin creating your tornado. The connection must be airtight for the experiment to work well.
2. Turn the connected bottles so that the bottle with the water is on top (like an hourglass).
3. Hold the bottles by the connected necks and rotate them quickly in a circular motion (clockwise or counterclockwise). Keep the bottles upright while rotating them.
4. A vortex will form in the top bottle as the water rushes through the hole into the lower bottle. You may need to repeat rotating the bottles before it works properly.

What Is Occurring

The vortex forms when centripetal force (inward force) and centrifugal force (outward force) are applied to a system with varying densities. In this case, the difference in density between air and water. Centripetal force also keeps the water spinning after you stop rotating the bottles. The vortex makes it easier for air to enter the top bottle and for water to pour into the lower bottle. The action of these forces is the same principle of how tornadoes form during a storm. Gravity is what pulls the water from the top bottle, which has lower air pressure, into the lower bottle, which has higher air pressure.

Name ______________________________ Date ______________ Period __________

Activity 16E

Tornado in a Bottle Variations

Version #1

In this version of the original tornado in a bottle experiment, you will add lamp oil to the water instead of the dish soap and glitter and predict the outcome. Use a type of cooking oil (with some color) if you cannot find lamp oil.

Materials

- 2 oz of lamp oil in addition to the materials used in the original tornado in a bottle experiment

Procedure

1. Repeat the original instructions but add the 2 oz of lamp oil to the water instead of the soap and glitter.
2. What do you think will occur when you create the vortex and the water flows into the lower bottle?

   ____

   ____

   ____

   ____

   A. Why do you think this will occur?

   ____

   ____

3. What happened when you created the vortex?

   ____

   ____

   ____

   A. Did this match your prediction?

   ____

   B. What caused the reaction of the oil and water vortex?

   ____

   ____

Version #2

In this activity, you will predict and measure the time it takes to empty a one-liter bottle of water. Work with a partner to help you record the time.

Materials

- activity worksheet
- bucket
- 1 one-liter bottle
- stopwatch or timer on phone
- water

Procedure

1. How long do you predict it will take to empty the bottle while holding it still (*not* squeezing the sides) over the bucket?

   A. Holding still: ______________________

   B. How long do you predict it will take to empty the bottle while you are rotating it to form a vortex?

   With vortex: ______________________

2. Fill the bottle to the top with water.
3. Hold the full bottle over the bucket and use your stopwatch or phone timer to measure the time. Refill the bottle and repeat this three times to validate your time. Indicate the average time.

   A. Time 1: ______________________

   B. Time 2: ______________________

   C. Time 3: ______________________

   D. Average: ______________________

4. Refill the bottle to the top with water.
5. Hold the full bottle over the bucket and move it in a clockwise or counterclockwise motion to form a vortex. Keep moving the bottle until it is empty. Refill the bottle and repeat this three times to validate your time. Indicate the average time.

   A. Time 1: ______________________

   B. Time 2: ______________________

   C. Time 3: ______________________

   D. Average: ______________________

6. Were the times you predicted accurate?

   ____

7. Was one method quicker than the other? If so, why do you think this is the case?

   ____

   ____

   ____

   ____

   ____

Name ______________________ Date ____________ Period ________

Activity 16F

Keeping a Weather Journal

Objectives

- Observe weather changes of a specific location.
- Improve graphing skills.

Materials

- worksheet
- internet access

Instructions

Over the course of 10 consecutive days, track the weather of a chosen location. Your teacher may assign the same dates for the class. Use the following worksheet to record your data. Note that all data for each day must be included to receive credit for each day. Create a line graph with a key after you complete the chart.

Weather Journal				
Location:				
Data Source:				
Day / Date	Weather	High Temperature	Low Temperature	Relative Humidity
1.				
2.				
3.				

Continued

Goodheart-Willcox Publisher

Weather Journal				
Location:				
Data Source:				
Day / Date	**Weather**	**High Temperature**	**Low Temperature**	**Relative Humidity**
4.				
5.				
6.				
7				
8.				
9.				
10.				

Goodheart-Willcox Publisher

Name ______________________________ Date ______________ Period __________

CHAPTER 17 Air Quality

Chapter Review

Carefully study the chapter and then answer the following questions.

1. _____ Which of the following may be found in ambient air?
   A. Water vapor and assorted gases
   B. Bacteria, viruses, and microscopic organisms
   C. Microscopic solids, such as dust and pollen
   D. All of the above.

2. _____ *True or False?* The introduction of a substance into air that has harmful effects on organisms and the environment is air pollution.

3. _____ Carbon monoxide and carbon dioxide from fossil fuel combustion are examples of _____.
   A. environmental pollutants
   B. natural pollutants
   C. secondary pollutants
   D. primary pollutants

4. _____ Pollutants formed through chemical reactions between primary pollutants and natural compounds in the air are _____.
   A. tertiary pollutants
   B. secondary pollutants
   C. radon pollutants
   D. natural pollutants

5. _____ Which of the following can be used to help maintain indoor air quality?
   A. Keeping gas-fueled appliances in good working order
   B. Installing carbon monoxide alarms and radon monitors
   C. Preventing mold and using dehumidifiers to control moisture
   D. All of the above.

6. _____ *True or False?* Indoor air contaminants may originate with the building's construction and insulation materials.

7. _____ *True or False?* Most of the chemical pollutants are natural chemicals released into the atmosphere.

8. _____ The negative effect of methane on the atmosphere is more than 20% higher than that of _____.

A. carbon monoxide (CO)

B. sulfur dioxides (SO_x)

C. carbon dioxide (CO_2)

D. nitric oxides (NO_x)

9. _____ The naturally released methane accounts for what percentage of the methane in the atmosphere?

A. 22%

B. 32%

C. 52%

D. 83%

10. _____ *True or False?* Methane is released through anthropogenic sources, including natural fuel production, landfills, and livestock operations.

11. _____ The greatest natural source of CO_2 is the _____.

A. volcanic eruptions

B. respiration of living organisms

C. ocean

D. rain forest

12. _____ The greatest anthropogenic source of CO_2 is _____.

A. chemical production

B. electric power generation

C. manufacturing processes

D. fossil fuel use

13. _____ Which of the following is an odorless, tasteless, and poisonous gas produced primarily through incomplete fossil fuel combustion?

A. Carbon dioxide

B. Carbon monoxide

C. Nitrogen oxide

D. All of the above.

14. _____ Nitrogen dioxides are highly reactive gases, aerosols, and acids that form when _____.

A. nitrogen combines with oxygen

B. nitric oxide combines with oxygen

C. nitrogen dioxide combines with chemicals or water

D. All of the above.

15. _____ *True or False?* Nitrogen dioxides from natural sources exceed that of nitrogen dioxides from anthropogenic sources.

16. _____ *True or False?* Nitrogen oxides react with other airborne chemicals to form ozone and particulate matter.

Name ____

17. _____ Acid rain is produced when water and other chemicals combine with _____.

A. nitrous and nitric acids

B. inert minerals

C. ozone

D. All of the above.

18. _____ What percentage of sulfur dioxides in the atmosphere are primarily from anthropogenic sources?

A. Less than 48%

B. 63% to 74%

C. About 85%

D. Almost 100%

19. _____ *True or False?* Sulfur oxides are one of the key ingredients in acid rain.

20. _____ Which of the following statements about VOCs is true?

A. They are volatized when products containing VOCs are used

B. They are essential components of many products

C. They are found indoors and outdoors

D. All of the above.

21. _____ Ground-level ozone _____.

A. protects Earth from ultraviolet rays

B. is caused by reactions with other pollutants

C. occurs more often on cool, cloudy days

D. All of the above.

22. _____ *True or False?* Smog can form in any climate but it is worse during warmer, sunnier weather, and in low-lying areas surrounded by hills.

23. _____ *True or False?* The Clean Air Act includes a section on noise abatement.

24. _____ Which of the following are effects of noise pollution?

A. Creates difficulty for birds and frogs to find mates

B. Frightens predators or prey away from a habitat

C. Changes the distribution of birds

D. All of the above.

25. _____ Which of the following is *not* a source of light pollution?

A. Light from the moon and stars

B. Advertising lights and streetlights

C. Sporting venues and commercial properties

D. All of the above.

26. _____ *True or False?* The Air Pollution Act of 1955 was the first federal legislation for controlling air pollution.

27. _____ The Clean Air Act was amended in 1965 to authorize the federal government to set standards for controlling pollutant emissions from certain vehicles. Which of the following is the amendment that authorized the government to set these standards?

A. Air Quality Act

B. Motor Vehicle Air Pollution Control Act

C. The Clean Air Act of 1965

D. The Clean Air Act for the Prevention of Significant Deterioration (of air quality)

28. _____ Which of the following enabled an increase in the government's investigation and enforcement of interstate air pollution source inspections?

A. Air Quality Act of 1967

B. Motor Vehicle Air Pollution Control Act

C. The Clean Air Act of 1965

D. The Clean Air Act for the Prevention of Significant Deterioration (of air quality)

29. _____ Which of the following government agencies was created in 1970 to implement the four programs initiated by the Clean Air Act of 1970?

A. NESHAPs

B. NSPS

C. NAAQS

D. EPA

30. _____ A reduction in emissions made to compensate for a greenhouse gas emissions made elsewhere is _____.

A. emission offsetting

B. cap-and-trade setting

C. emission trading

D. All of the above.

31. _____ The amount of greenhouse gases emitted by something during a given amount of time is called a(n) _____.

A. emission print

B. carbon footprint

C. compliance print

D. offset print

32. _____ *True or False?* A zero emission vehicle does not emit pollutants through its exhaust.

33. _____ *True or False?* A person who chooses to reduce or eliminate activities that emit or cause harmful emissions can become carbon-neutral.

34. _____ A process in which carbon dioxide is removed from the atmosphere and held in solid or liquid form in the environment is called carbon _____.

A. offsetting

B. emission control

C. reduction

D. sequestration

Name ____

Vocabulary Review

Match the following terms with the appropriate definition or description.

A. active sampling
B. Aerostat/Flyer
C. Air Quality Index
D. automatic sampling
E. black carbon monitor
F. Geospatial Measurement of Air Pollution
G. passive sampling
H. Sensor NeTwork Intelligent Emissions Locator
I. SPod

1. _____ Measures the five major air pollutants regulated by the Clean Air Act: particle pollution, ground-level ozone, carbon monoxide, nitrogen dioxide, and sulfur dioxide.
2. _____ A simple method that uses diffusion tubes or discs that are open at one end and have a chemical or biological absorbent at the other end.
3. _____ The collection of samples through physical or chemical means, such as pumps that move air through a filter or a chemical collector.
4. _____ A sophisticated and expensive method that uses measurement devices placed in multiple sites and monitored in one location.
5. _____ Collects real-time data using Google street view vehicles equipped with high time-resolution air pollution instrumentation.
6. _____ Uses monitors coupled to digital billboards to reduce costs by using the billboard's electricity and internet access.
7. _____ A solar-powered device that can be placed in areas without an available power source.
8. _____ A short-lived climate forcer that serves as an indicator of combustion emissions.
9. _____ Used for sampling hard-to-sample lofted plumes of wildland fires and prescribed burns.

Thinking Critically

1. What are the health and environmental concerns for animals, such as birds, insects that pollinate at night, and sea turtle hatchlings, that are caused by light pollution?

____

____

____

____

2. What is an airshed? How is it defined?

____

____

____

____

3. Briefly explain what cap-and-trade is and how it works.

4. How effective are air purifiers, such as ozone generators and ionizers, at cleaning indoor air? Do they do more harm than good? Explain your answer.

5. How much responsibility do you think governments and industries have for keeping air quality healthy?

6. What changes would you propose to reduce air pollution if you were a sustainability expert representing the needs of your community to local government?

7. During the summer of 2023, the United States experienced the worst air quality in its history due to changes in wind currents, existing weather fronts, and hundreds of uncontrolled wildfires burning in Canada. How could the smoke from these, and other wildfires, impact public health throughout the globe?

Name ______________________ Date ______________ Period __________

Activity 17A

Technology Highlight

Objectives

- Identify new technologies being used to benefit conservation and/or sustainability.
- Prepare and print a three-fold pamphlet.

Materials

- internet access
- computer and printer

Instructions

Select an emerging natural resources-related technology. Research and determine how the technology was developed and what it can do to benefit conservation and/or sustainability. Create a three-fold, color informational pamphlet explaining the role of the technology. Include applicable images. The pamphlet should answer the following questions. Present your pamphlet to the class.

1. How does this new technology work?

____
____
____

2. Who developed this new technology?

____
____
____

3. Why was this technology needed?

____
____
____
____

4. What is this technology designed to improve or replace?

____
____
____
____

5. What are the advantages of using this technology?

6. What are the drawbacks to using this technology?

Name ______________________________ Date ______________ Period ________

Activity 17B

International Airsheds

Objectives

- Improve geographical knowledge.
- Understand how wind currents can carry radiation and other contaminants.

Materials

- map of Europe
- world map
- activity worksheet

The Explosion at Chernobyl

On April 26, 1986, the Chernobyl nuclear power plant, which is near the city of Pripyat, in northern Ukraine, became the site of one of the world's worst nuclear power plant disasters. Plant operators were performing tests to develop a safety procedure for power outages that would keep the core cool until power was restored or the back-up generators provided power. The operator failed to follow procedure, which caused unstable conditions that resulted in an uncontrolled nuclear chain reaction. The cooling water was vaporized by the energy surge, and caused a steam explosion that ruptured the reactor. The blast blew radioactive gases and particles as high as three miles into the atmosphere. The exposed reactor core and surrounding debris, such as graphite, burned for ten days after the explosion, and continued releasing radioactive contamination.

Two plant workers were killed by the explosion and 29 others later died from radiation exposure. Within days, more than 120,000 people were evacuated from a twenty five-mile radius around the plant.

Where It Went

The airborne gases and particles from the explosion were spread by the winds. The contaminants were carried northwest toward Poland and Scandinavia, and then southwest across central Europe. The contamination also moved east across Asia, over Japan and the North Pacific, and eventually reached the west coast of North America. As the reactor continued to burn, it released radiation that moved south and west of the plant.

Effects of the Explosion

The first few weeks following the Chernobyl blast were filled with confusion. Some European countries ordered the destruction of millions of dollars of contaminated produce, milk, and livestock. In other European countries, people were told that it was safe to consume these products. Farmers suffered huge financial losses when countries in other parts of the world refused to import produce from Europe. Radiation, even at low levels, can increase incidence of cancer. The digestive system, blood pressure, and heart are particularly sensitive. The explosion also strained relations between the Soviet Union and other nations. Many countries were angered by the Soviet Union's delay in reporting the accident—officials announced it days later on April 29, 1986.

Today, the damaged reactor is entombed in thick concrete and a safe, steel containment shield. But the disaster is still taking its toll. Some scientists predict that within the next few decades, thousands of people who were exposed to the radiation could develop cancer as a result of their exposure.

Instructions

In this activity, you will be marking the areas outside of the Ukraine that experienced different levels of fallout from the Chernobyl nuclear disaster. Draw a dot on each of the cities and then use a ruler to draw lines connecting the dots from the power plant to the city. Use the color specified for each day.*

1. **Day 1, April 26 (red)**

   The number four reactor at Chernobyl nuclear power plant explodes and ignites. The prevailing winds were from the south and east. Draw a star to indicate the plant's location.

2. **Day 2, April 27 (dark blue)**

   Winds blow a radioactive cloud northwest over Gdansk, Poland.

3. **Day 3, April 28 (dark green)**

   Radiation reaches Stockholm, Sweden; Helsinki, Finland; and Oslo, Norway.

4. **Day 4, April 29 (brown)**

   Radiation continues moving north through Scandinavia and reaches Trondheim, Norway; Copenhagen, Denmark; and Prague, Czechoslovakia.

5. **Day 5, April 30 (orange)**

   Radiation moves over Munich, West Germany; Vienna, Austria; and Geneva, Switzerland.

6. **Day 6, May 1 (black)**

   Radiation reaches Rome, Italy; Budapest, Hungary; Zagreb, Yugoslavia; Paris, France; and Tromso, Norway.

7. **Day 7, May 2 (yellow)**

   Small amounts of radiation are measured near Reykjavik, Iceland.

   Radiation reaches Sofia, Bulgaria; Bucharest, Romania; Brussels, Belgium; London, England; and high amounts fall with the rain north of London.

8. **Day 8, May 3 (purple)**

   Radiation reaches Glasgow, Scotland; Athens, Greece; and Ankara, Turkey.

9. **Day 9, May 4 (pink)**

   Radiation reaches Beirut, Lebanon.

10. **Day 10, May 5 (light green)**

    Radiation detected in Damascus, Syria.

    NOTE

    The following cities are only included on the world map.

11. **Day 11, May 6 (light blue)**

    Radiation reaches Kuwait, the capital of Kuwait and Xian, China.

12. **Day 12, May 7 (red-orange)**

    Radiation reaches Tokyo, Japan.

13. **Day 18, May 13 (yellow-green)**

    Slight amount of radiation is detected in Richland, Washington, in the United States.

*The days are estimations of the movement of radiation fallout from the disaster.

Name ____

Goodheart-Willcox Publisher

Andrei Minsk/Shutterstock.com

Name ______________________________ Date ______________ Period __________

CHAPTER 18 Interactions

Chapter Review

Carefully study the chapter and then answer the following questions.

1. _____ The effect that organisms living in the community have on each other in an ecological community is a(n) _____.
   A. confrontation
   B. interaction
   C. edge effect
   D. distraction

2. _____ Which of the following statements about ecological communities is *not* true?
   A. It is composed of populations of different species
   B. There is often greater biodiversity at the edge habitats
   C. The boundaries are well-defined and permanent
   D. The community structure changes at the boundaries

3. _____ *True or False?* Ecosystems with abundant resources typically have a low amount of biodiversity.

4. _____ *True or False?* The habitat that is preferred by a species is its native habitat.

5. _____ Species with limited habitat _____.
   A. live in ecosystems that often require protection
   B. are often low in number
   C. are less likely to adapt to different habitats
   D. All of the above.

6. _____ *True or False?* How a species lives in its habitat and the role that it plays is its ecological niche.

7. _____ Interactions between different species are _____.
   A. intraspecific
   B. interspecific
   C. habitual
   D. adaptations

8. _____ Interactions between members of the same species are _____.
   A. intraspecific
   B. interspecific
   C. habitual
   D. adaptations

9. _____ *True or False?* Each living organism in an ecological community is exclusively part of one food chain.

For questions 10–20, match the components of an ecosystem with the proper location in an ecosystem.

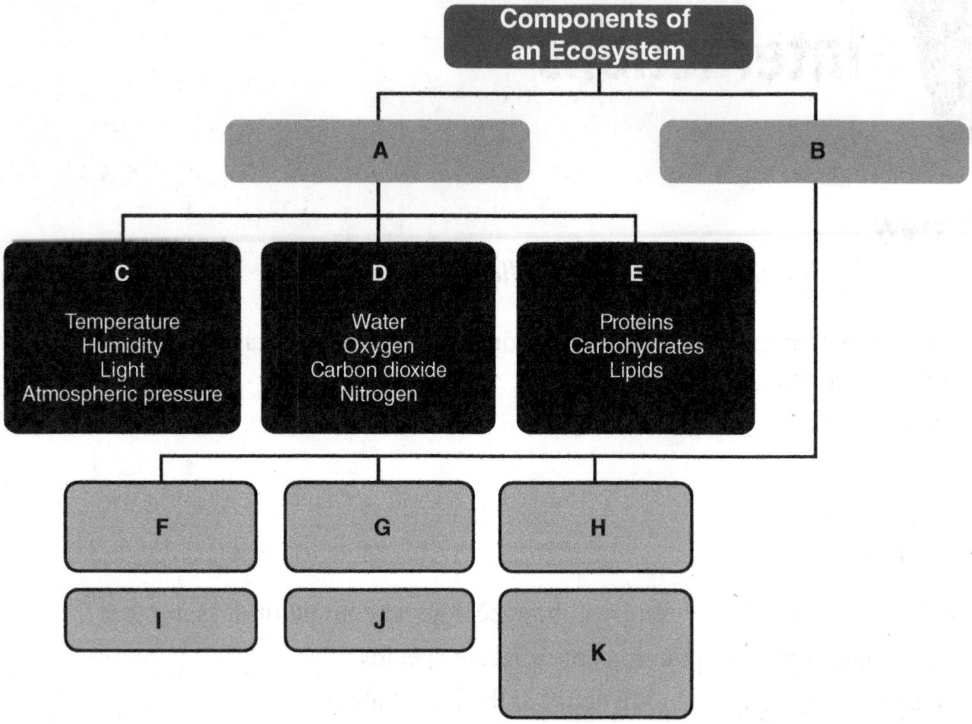

Goodheart-Willcox Publisher

10. _____ Physical factors
11. _____ Producers (green plants)
12. _____ Autotrophs
13. _____ Abiotic components
14. _____ Heterotrophs
15. _____ Biotic components
16. _____ Organic substances
17. _____ Consumers (animals)
18. _____ Inorganic substances
19. _____ Decomposers (microorganisms)
20. _____ Sapotraophs, detrivores, detritus feeders

Name ____

21. _____ *True or False?* Approximately 10% of the original energy in a food chain is passed on as the energy flows through each trophic level.

22. _____ A species on which others in the community rely and whose removal would change the ecosystem drastically is a(n) _____.

A. avian species

B. aquatic species

C. keystone species

D. vertebrate species

23. _____ The great white shark, saltwater crocodile, lion, grizzly bear, and wolf are _____.

A. species on which others rely

B. predator keystone species

C. apex predators

D. All of the above.

24. _____ *True or False?* Some keystone predators are not apex predators.

25. _____ *True or False?* Herbivores, such as elephants and prairie dogs, are not keystone species.

26. _____ *True or False?* A mutualistic keystone species is a species on which another species depends and cannot survive if it is removed from the ecosystem.

27. _____ Species that are regularly observed to determine if an ecosystem is at risk are _____.

A. foundation species

B. flagship species

C. umbrella species

D. indicator species

28. _____ A species that plays a strong role in establishing structure in a community is a _____.

A. foundation species

B. flagship species

C. umbrella species

D. indicator species

29. _____ Species that serve as ambassadors for environmental causes are _____.

A. foundation species

B. flagship species

C. umbrella species

D. indicator species

30. _____ A species that can help tell the overall health of an area, such as a forest, is a _____.

A. foundation species

B. flagship species

C. umbrella species

D. indicator species

Vocabulary Review

Match the following terms with the appropriate definition or description.

A. adaptation
B. commensalism
C. competition
D. competitive exclusion
E. evolution
F. herbivory
G. mutualism
H. natural selection
I. neutralism
J. parasitism
K. predation
L. symbiotic relationship

1. _____ An interaction in which one or both organisms benefit, or one suffers due to the other organism's actions.
2. _____ When the relationship is beneficial to one species and has no effect on the second species.
3. _____ When two species interact for mutual benefit.
4. _____ When an interaction may be detrimental to both species.
5. _____ Occurs when two species compete for the same resources but one is better suited to using the niche.
6. _____ The process in which only the organisms best adapted to an environment survive and transmit their genetic characteristics to their offspring.
7. _____ The process of change in which an organism becomes better suited to live and reproduce in its environment.
8. _____ The gradual change in genetics of a population over many generations.
9. _____ When one organism lives off another and causes it harm.
10. _____ An interaction that is beneficial to one species and deadly to the second.
11. _____ An interaction in which plant materials are consumed by another organism.
12. _____ Occurs when neither species is affected.

Match the following organisms with the appropriate definition or description.

A. apex predators
B. autotrophs
C. carnivores
D. decomposers
E. herbivores
F. heterotrophs
G. omnivores
H. primary consumers
I. primary producers
J. secondary consumers

13. _____ Living creatures that reduce organic matter into simple nutrients and return the nutrients to the soil.
14. _____ Autotrophic organisms, such as plants, algae, and some bacteria that are at the base of the trophic hierarchy.
15. _____ Organisms that gain energy from the sun and use photosynthesis to produce their own food.

Name ______________________________

16. _____ Organisms that consume nutrients but do not produce them.

17. _____ The organisms at the second level of the trophic hierarchy.

18. _____ Carnivores at the top of the trophic hierarchy in an ecosystem that have no natural enemies.

19. _____ An organism that eats only plants.

20. _____ The predator species at the third level of the trophic hierarchy who eat herbivores.

21. _____ Animals that eat other animals as the majority of their diet.

22. _____ Animals that eat both plants and animals.

Thinking Critically

1. What are the three characteristics that a species must have to be considered a keystone species?

____
____
____

2. Why has the reintroduction of wolves to Yellowstone National Park been met with opposition?

____
____
____

3. Which FWS programs have been established to control invasive species?

____
____
____

4. Why are there more herbivores than carnivores?

____
____
____

5. Why are there no more than four or five links in a food chain?

____
____
____

6. Why is a food web a better description of feeding relationships than a food chain?

____
____
____

Notes

Name ______________________ Date ____________ Period ________

Activity 18A

Food Chains and Food Webs

Objectives

- Understand how organisms align in a food web.
- Identify the roles of producers, primary consumers, and secondary consumers.

Material

- activity sheet

Instructions

Food chains show the relationship between producers (plants), primary consumers (animals that eat plants), and secondary consumers (animals that eat other animals).

Example: seeds → sparrow → hawk → fox
A food web is a series of interconnected food chains.
Example:

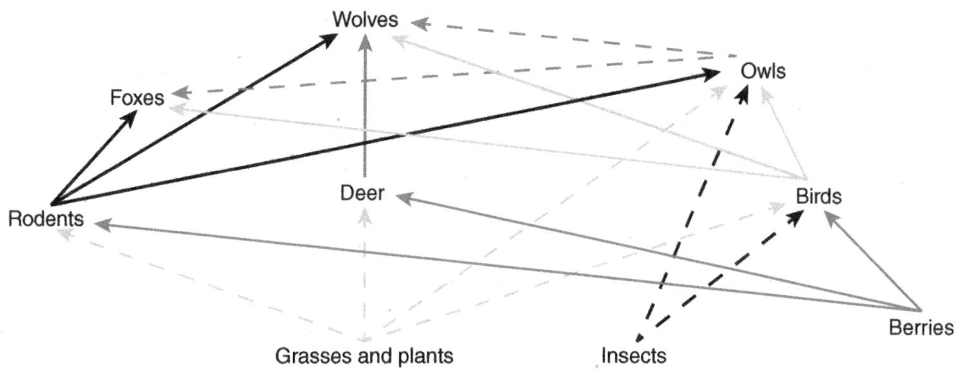

Goodheart-Willcox Publisher

1. Create five food chains, each with at least four organisms each. Each chain must include at least one producer, one primary consumer, and one secondary consumer.

   A. Food chain 1: ______________________

   B. Food chain 2: ______________________

   C. Food chain 3 ______________________

   D. Food chain 4: ______________________

   E. Food chain 5: ______________________

2. Draw a food web with at least 15 different organisms.

Notes

Name ______________________ Date ______________ Period __________

Activity 18B

Food Webs

Objectives

- Understand the interrelatedness of food webs.
- Explain how populations affect other populations.
- Explain what may occur when an invasive species enters a food web.

Materials

- hole puncher
- index cards
- marker
- masking tape or painter's tape or outdoor chalk
- scissors
- yarn

Preparation

1. Decide on a biome and write a list of animals, plants, and other organisms that live in that biome. Review the list as a class and eliminate organisms that do not live in your chosen area. You must include the sun, and organisms from different trophic levels, including microscopic bacteria.
2. Write the name of each organism on a separate index card and write the number that represents the organism's trophic level.
   - 1 primary producers
   - 2 primary consumers
   - 3 secondary consumers
   - 4 tertiary consumers
   - 5 quaternary consumers
3. Punch holes in the corner of each card and have students use a piece of yarn to hang the cards around their necks. You may want to assign one or more students to be in charge of the yarn and the scissors, and others to help determine which organisms should be connected.
4. If you are doing this activity outside, you can make five lines (to represent the trophic levels) in parallel that are about 5′ long and 2′ apart. If you are working inside, use the masking tape to make the lines on the floor. Label the lines one through five.

Instruction

1. Have the sun stand before the first line. The students should sit on the line that matches their number.
2. A ball of yarn will be passed from the sun, to the primary producers, and so on. Once the yarn reaches the end of the chain, it is cut and you will create another chain. For example, a steer eats grass, which gets its energy from the sun. The sun passes the yarn to the grass, who then passes it to the steer. When it reaches the apex predator, the apex predator dies and it goes back to any scavengers and then to the decomposers.

3. Stand up when the yarn is passed to you and loosely wrap the yarn around your finger before you pass it on to the next organism.
4. When the yarn reaches the apex predator, cut it and give the ball of yarn to the sun. You can give the dead organism a brown or black thread to pass back through the web to indicate how dead organisms fit into the web.
5. Continue creating chains and building the web as time and participants allow. When you are all connected, raise your hands in the air so everyone can see the complexity of the web.
6. What would happen if an organism were removed from the chain? Choose one person to pull on the strings that he or she holds. Anyone who feels the tug is directly affected by that organism.
7. Anyone who felt a tug should lightly tug on their strings. Once again, anyone who feels the tug is directly affected by those organisms. What should happen if the sun pulls on the strings he or she is holding?

Extending the Activity

1. Choose two or three organisms to go extinct and have them drop their yarn. Who else is affected when these species are removed?
2. What would happen if an invasive species entered the web? Choose one organism to be outcompeted by an invasive species.
3. Have everyone hold their hands up in the air and then have someone cut the strands for each thing that the organism eats. What happens to the web? Have everyone who was connected to the organism (before the strings were cut) sit down.

Name ______________________________ Date ______________ Period __________

CHAPTER 19 Endangered Species

Chapter Review

Carefully study the chapter and then answer the following questions.

1. _____ Species extinction occurs naturally at a background extinction rate of _____.
   A. one to five species per year
   B. one to five species per decade
   C. 1000 species per year
   D. 1000 species per decade

2. _____ The environmental organization dedicated to global conservation and the protection of species and their habitats is the _____.
   A. MBTA
   B. MMPA
   C. IUCN
   D. ESA

3. _____ Which of the following criteria is used by the IUCN to determine a species' status?
   A. Probability of extinction
   B. Population size and population restrictions
   C. Population reduction rate and geographic range
   D. All of the above.

4. _____ According to the USFWS, a species that is in danger of extinction throughout all or a significant portion of its range is classified as _____.
   A. threatened
   B. endangered
   C. near threatened
   D. excluded from the wild

5. _____ According to the USFWS, a species that is likely to become endangered in the foreseeable future throughout all or a significant portion of its range is classified as _____.
   A. threatened
   B. endangered
   C. near threatened
   D. excluded from the wild

6. _____ *True or False?* A species that is endangered can receive protection under the Endangered Species Act while waiting to be added to the federal list.

7. _____ A species that is locally extinct and no longer found in an area it used to inhabit but is still found elsewhere has experienced _____.

A. limitation

B. degradation

C. extirpation

D. genetic variation

8. _____ *True or False?* Habitat degradation changes an area in such a way that it can no longer meet the needs of its inhabitants.

9. _____ Which of the following occurs with habitat fragmentation?

A. Migration patterns are interrupted

B. There is less living space for many animals

C. Feeding grounds are separated

D. All of the above.

10. _____ *True or False?* The genetic variation available to a species is minimally affected when the population declines.

11. _____ *True or False?* Young, unhealthy, or old animals are more vulnerable to disease.

12. _____ Wildlife diseases that can infect human beings are referred to as _____.

A. feral diseases

B. specialized diseases

C. zoonotic diseases

D. All of the above.

13. _____ *True or False?* Animal species with long gestation periods build populations more quickly than those with short gestation periods.

14. _____ The lowest number of individuals needed for the species to survive in the wild is its _____.

A. reproduction limit

B. minimum viable population

C. maximum viable population

D. ideal viable population

15. _____ Overhunting and overfishing endangers species because _____.

A. the genetic variation is reduced

B. it may begin experiencing genetic defects in offspring

C. the number of breeding animals is reduced

D. All of the above.

16. _____ *True or False?* Studies indicate that it is not possible to have a sustainable, productive ecosystem without biodiversity in the landscape.

Name ______________________________

17. _____ *True or False?* A transect line is marked at regular intervals and used to measure the square feet of an area.

18. _____ Species of microorganisms, insects, fungi, and plants whose presence or absence are used to assess the quality of an environment are _____.

A. bioindicators

B. rare species

C. captivators

D. All of the above.

19. _____ *True or False?* Captive breeding programs are used to breed and reintroduce endangered animals into the wild.

20. _____ Which of the following statements about species survival plans is accurate?

A. It is *ex situ* conservation

B. Used to breed animals with genetic diversity

C. Keep some specimens in captivity

D. All of the above.

For questions 21–27, match the acts of legislation to the appropriate description.

A. The Lacey Act of 1900 and 2008

B. The Migratory Bird Treaty Act (MBTA) of 1918

C. Migratory Bird Conservation Act of 1929

D. Migratory Bird Hunting and Conservation Stamp Act (1934)

E. Marine Mammal Protection Act (MMPA) of 1972 and 1994

F. Endangered Species Act (ESA), 1973

G. Wild Bird Conservation Act (WBCA) of 1992

21. _____ Anyone over 16 must purchase a stamp to hunt for waterfowl. The proceeds are used for wildlife conservation.

22. _____ Requires special permits for the importation of wild bird species for zoos, recreation, or personal pets. This Act is intended to protect birds at their point of origin.

23. _____ Created the United States Migratory Bird Conservation Commission (MBCC) that approved purchase of open land or water areas by the US Fish and Wildlife Service to assist birds in migration.

24. _____ It is illegal to hunt, trap, or kill migratory birds as they move between the United States and Canada.

25. _____ It is unlawful to import, export, sell, trade, or use any fish, plant, or other wildlife listed in the Act.

26. _____ Protects all marine mammals within United States' waters.

27. _____ Works with both domestic and international conservation to prevent the killing of species listed as endangered or threatened.

Vocabulary Review

Match the CRITICAL classifications to the appropriate description.

A. critically endangered species (CR)
B. data deficient (DD)
C. endangered species (EN)
D. extinct in the wild species (EW)
E. extinct species (EX)
F. least concern (LC)
G. near threatened species (NT)
H. vulnerable species (VU)

1. _____ Species with no living specimens.
2. _____ Species with some specimens living in captivity but with no specimens living in the wild.
3. _____ Species facing an extremely high risk of becoming extinct.
4. _____ Species facing a high risk of extinction in the wild.
5. _____ Species likely to become endangered unless action is taken to identify and correct factors that will likely lead to its demise.
6. _____ Species that may face population decline in the future and should be monitored for changes.
7. _____ Species that are doing well and have no cause for concern.
8. _____ There is inadequate information on abundance or distribution to make an assessment about a species' status.

Match the methods used to identify, track, and count the individuals with the appropriate description.

A. direct observation *in situ*
B. DNA barcoding
C. incomplete count
D. indirect count
E. mark-recapture
F. quadrat analysis
G. random sampling
H. remote sensing
I. stratified sampling
J. systematic sampling
K. total count

9. _____ A method of collecting information about organisms or environments from a distance using aircraft, satellites, or drones.
10. _____ A method used to estimate a species population and determine its distribution over a random area using a set of similar shapes.
11. _____ Every member of a species in an area is counted.
12. _____ Accomplished by taking samples of an area according to features that subdivide the overall area.
13. _____ Counts are done in specified areas that use the average of specimens counted in multiple areas.
14. _____ Performed by randomly selecting and sampling small areas within the larger area.
15. _____ Requires capturing, marking, releasing, and recapturing animals during repeated samplings.

Name ______________________________

16. _____ Requires researchers on the ground to manually count the animals.

17. _____ Sampling in a systematic way, such as on a grid system.

18. _____ The researcher uses indirect signs of the species' presence and does not count the animals themselves.

19. _____ Used to identify known species by comparing their barcode DNA sequences to reference sequences in public data bases.

Thinking Critically

1. How does the introduction of an invasive species cause native species to become endangered?

____
____
____
____

2. Why is a species with a limited diet and exclusive habitat more vulnerable to extinction?

____
____
____
____

3. What is the importance of private lands to endangered species?

____
____
____
____

4. Is the Endangered Species Act detrimental to some species?

____
____
____
____

5. What is the Endangered Species Reserve Program? How would it work in practice?

____
____
____
____

Notes

Name ______________________________ Date ______________ Period __________

Activity 19A

Persuasive Speech Outline

Objectives

- Gather and organize information in preparation for a speech.
- Develop organization and preparation skills.
- Identify the advantages and/or disadvantages of the Endangered Species Act.
- Making a good impression and leaving your audience forming opinions on the validity of the Endangered Species Act.

Materials

- activity sheet
- internet access

Instructions

This outline activity is designed to help you gather the information that you will be including in a persuasive speech. Your goal is to persuade your audience to believe in your cause and hopefully take action. Using an outline will help you organize your thoughts and ideas and guide you logically in your speech. As you develop your outline and research your topic, you will also gather additional information that can be useful in answering questions from the audience.

Choose a Topic

This speech is about an environmental issue dealing with endangered species and how you stand on the issue. The more specific your topic, the better. Carefully consider potential arguments on the topic or use a topic assigned by your teacher.

1. Speech Topic: ______________________________
2. Speech Title: ______________________________
3. It is important that you know how you stand on the topic and why you feel that way *before* you begin writing because it will help you put more feeling into your speech. Are you for or against the Endangered Species Act as it stands today? Briefly explain why you feel that way.

   ____

   ____

   ____

   ____

   ____
4. Sum up your argument in a one-sentence, definite statement.

   ____

   ____

Identify Your Audience

It has been said that writing a speech involves meeting the expectations of the people in your audience, whether it is to motivate, entertain, or persuade. Not only should you know who your audience is, but you should know why they are present and what they want to gain from your speech.

1. Who is your intended audience? ______________________________

   A. Why are they present?

   ____

   ____

   ____

   B. What do they want to take away from your speech?

   ____

   ____

   ____

2. How much about your topic do you think your audience knows?

   ____

   ____

   ____

3. What do you want your audience to be able to take away from the speech? What is the thing you want them to remember or how do you want to persuade them?

   ____

   ____

   ____

   ____

4. What are the challenges you think you will have with this audience? How will you overcome those challenges?

   ____

   ____

   ____

Section One: The Introduction

1. What are you going to use to catch the audience's attention? Be specific and tell exactly what you are going to do for your attention grabber. Share a shocking fact or statistic or open with a question to get listeners engaged early.

   ____

   ____

   ____

   ____

Name ____

Section Two: The Body

Use the body to summarize your points and explain why your subject is of dire need. Repetition is very useful in a persuasive speech because it leaves an impression on the audience. Repeat critical points, key words, phrases, and themes. You also need to use transitional phrases that signal intent and lead the audience to the next point. Rhetorical questions, for example, are often used for making a transition. Saying "what does this mean," followed by a pause can get attention and create anticipation. The lack of transitional phrases can make the speech awkward and difficult to follow.

1. What are your main points? (Identify at least three)

   A. Section 1: ____

   ____

   ____

   B. Section 2: ____

   ____

   ____

   C. Section 3: ____

   ____

   ____

   D. Section 4: ____

   ____

   ____

   E. Section 5: ____

   ____

   ____

Section Three: The Conclusion

Ending your speech with strong closing lines is important if you want your audience to take action. Your closing lines should give the audience the specifics of what it is you want them to do and the benefits of doing so. A general rule of thumb is to include about 5% to 10% of your speech in the conclusion and to not add more information. The transition to the end of your speech should be smooth and help drive the message home.

1. What ideas do you have for concluding your speech? Metaphors, analogies, surprise, and axioms can work if they are built up and placed in the best spot, such as near the end.

   ____

   ____

   ____

   ____

   ____

   ____

2. How do you plan to "tell them what you told them"? This is where you recap your thesis and tie it all together.

____

____

____

____

____

____

Gather the Facts

1. List the facts that you will be using for each of the sections. List at least three facts per section and include a source for each fact.

Section 1: The Introduction

1. Fact 1: ______________________________

   Source: ______________________________

2. Fact 2: ______________________________

   Source: ______________________________

3. Fact 3: ______________________________

   Source: ______________________________

Section 2: The Body

1. Fact 1: ______________________________

   Source: ______________________________

2. Fact 2: ______________________________

   Source: ______________________________

3. Fact 3: ______________________________

4. Fact 4: ______________________________

   Source: ______________________________

5. Fact 5: ______________________________

   Source: ______________________________

6. Fact 6: ______________________________

   Source: ______________________________

Name ____

Section 3: The Conclusion

1. Fact 1: ____
   Source: ____
2. Fact 2: ____
   Source: ____
3. Fact 3: ____
4. Fact 4: ____
   Source: ____

Write Your Speech!

Use the information you have gathered to guide the development of your speech. Bring the facts for each section together. Use connections to your life or your audience's experiences to make the content more meaningful.

- The speech must be typed
- The speech must be at least 2 1/2 pages (double-spaced)
- The speech should be approximately 5 minutes
- Use proper spelling and punctuation
- Read through and rewrite
- Add notes to indicate emphasis and pauses

Practice Out Loud!

Practicing your speech out loud in the same manner as you would for your presentation will help you give a successful presentation. Speaking as though you are presenting will help you with the following:

- Determine the length of your speech and whether it is under or over the limit
- Identify awkward wording and prevent the creation of tongue twisters
- Identify sentences that are too long or possibly too short
- Identify words that should be emphasized and places to pause
- Speak with conviction and emotion

Name ______________________ Date ____________ Period ________

Activity 19B

Captive Breeding Research Assignment

Instructions

Students can work alone or in groups of two.

Select an animal from the list of endangered animals on the website of the USFWS, World Wildlife Fund, or IUCN Red List. Each group must have a different topic.

Develop a "sales pitch" to get people to support your endangered species. A sales pitch is a condensed presentation where a salesperson (you) explains the nature and benefits of their product (the animal you selected), ideally in one or two minutes. Explain why people should give their resources to this species in a captive breeding program. Your pitch can be a poster, a video, a PowerPoint, or anything to make the students in the class care about your species. The presentation must sell your animal to the class!

Include the Following in Case Studies:

- Species name, natural range, habitat, and basic information
- A map of the species' historic and current ranges and/or historic and current population statistics
- Reason the species is threatened or endangered
- When the captive breeding program began
- Difficulties with maintaining the wild population and with breeding in captivity
- Assessment of why the captive breeding program has been successful or not
- Explanation of how the program might help the overall biodiversity of the regions where the species naturally lives

Make a Choice

- Once all pitches have been made, the class will figure out which species gets public interest.
- Teams cannot financially support their own creature.
- Each student can support (up to) three captive breeding programs.
- Each student will be given 3 Post-it-Notes.
  - Blue notes = $1 million
  - Yellow notes = $2 million
  - Pink notes = $3 million
- Each student will select which program receives all or some of the support (this is a decision per person, not per group).
- The three groups with the most donations will receive prizes at the end, so be creative; lives are depending on you!

Name ______________________________ Date ______________ Period __________

CHAPTER 20

Fisheries

Chapter Review

Carefully study the chapter and then answer the following questions.

1. _____ Hatcheries for fish and invertebrates are production facilities focused on _____.
   A. rebuilding endangered species
   B. breeding, nursing, and rearing aquatic species
   C. rebuilding oyster reefs and wild populations
   D. All of the above.

2. _____ *True or False?* Commercial marine fisheries are located primarily along coastal areas.

3. _____ NOAA (no·ə) is an agency within the Department of Commerce that performs which of the following duties?
   A. Conducts research to improve environmental stewardship
   B. Provides climate monitoring to fisheries management
   C. Focuses on the conditions of the oceans, waterways, and atmosphere
   D. All of the above.

4. _____ The National Marine Fisheries Service is responsible for which of the following?
   A. National weather bulletins and foul weather warnings
   B. Policing of US waters in the Gulf of Mexico
   C. Stewardship of resources within the US EEZ
   D. All of the above.

5. _____ Each region of the US fisheries has its own _____.
   A. exclusive fishing zones
   B. fish management plan
   C. restrictions on whaling
   D. All of the above.

6. _____ Which of the following agencies is responsible for regulating and overseeing US fish hatcheries?
   A. US Coast Guard
   B. National Oceanic and Atmospheric Administration
   C. Fish Management Plans Agency
   D. US Fish and Wildlife Service

7. _____ Protective laws are strictly enforced in marine fisheries by which of the following agencies?
   A. National Oceanic and Atmospheric Administration
   B. US Coast Guard
   C. State and local governments
   D. All of the above.

8. _____ Harvesting a stock at a rate higher than the rate that produces its maximum sustained yield is considered _____.
   A. overregulated
   B. overfishing
   C. overfished
   D. endangered

9. _____ A stock having a population size that is too low and that jeopardizes the ability to produce its maximum sustained yield is considered _____.
   A. overregulated
   B. overfishing
   C. overfished
   D. endangered

10. _____ Which of the following is fishing as either a means of survival or to preserve cultural traditions?
   A. Commercial fishing
   B. Subsistence fishing
   C. Sport fishing
   D. Recreational fishing

11. _____ Which of the following is harvesting fish for profit?
   A. Commercial fishing
   B. Subsistence fishing
   C. Sport fishing
   D. Recreational fishing

12. _____ Which of the following is done for pleasure and competition?
   A. Subsistence fishing
   B. Commercial fishing
   C. Trawling for fish
   D. Sport fishing

13. _____ Which of the following is aquatic wildlife that is inadvertently caught with commercial harvesting?
   A. Dredges
   B. Longliners
   C. Bycatch
   D. Drifted catch

Name ____

For questions 14–21, match the commercial fishing methods with the appropriate description.

A. bottom longlines
B. dredge
C. drifting longline
D. harvesting machine
E. hydraulic dredge
F. pole-and-line fishing
G. seines or purse seine
H. trawling

14. _____ A commercial fishing method that uses nets with funnel-shaped bags towed behind the vessel

15. _____ Uses teeth to dig mollusks and damages much of the catch as well as killing excessive bycatch

16. _____ An indiscriminate fishing method that uses a long main line with hundreds or thousands of baited hooks on branch lines

17. _____ A method that is limited to harvesting small organisms

18. _____ A method that uses a floating keg and lighted flag buoy to mark its position and hangs above the sea floor to prevent other creatures from eating the bait

19. _____ A commercial method that uses multiple hooks on a single line

20. _____ A fishing method that uses nets attached to a buoy and to a vessel, which travels in a wide circle laying the ropes and returning to the buoy where the net is lifted onto the ship

21. _____ A method that uses jets of water under high pressure to wash mussels from the ocean floor

22. _____ The estimated number of people employed in direct fishing and related jobs is _____.
A. 19 million
B. 59 million
C. 79 million
D. 109 million

23. _____ According to NOAA, in 2020, US commercial and recreational fishing supported over _____.
A. 1.7 million jobs
B. 2.7 million jobs
C. 27 million jobs
D. 207 million jobs

24. _____ Which of the following pose threats to fisheries?
A. Ghost fishing and invasive species
B. Global climate change
C. Overfishing and bycatch
D. All of the above.

25. _____ Global climate change is endangering fisheries through which of the following?
A. Rising sea levels
B. Higher ocean temperatures
C. Increased acidification
D. All of the above.

For questions 26–35, identify the indicated parts of fish anatomy.

26. _____ vent

27. _____ anal fin

28. _____ nares

29. _____ dorsal fin

30. _____ gills

31. _____ pectoral fin

32. _____ lateral line

33. _____ caudal fin

34. _____ operculum

35. _____ pelvic fin

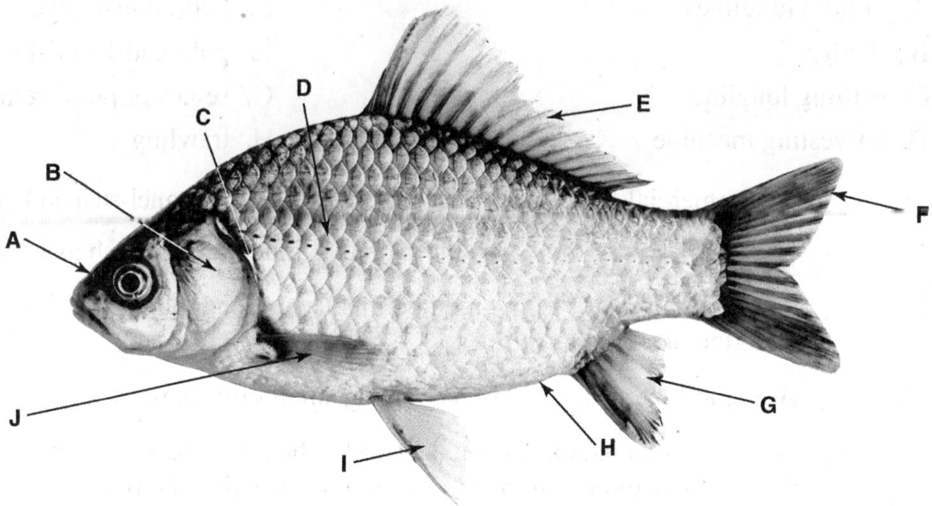

Nosyrevy/Shutterstock.com

36. _____ Which of the following is a location producing at optimal production but that cannot be expected to grow?
   A. Sustainable fishery
   B. Fully exploited fishery
   C. Overexploited fishery
   D. Collapsed fishery

37. _____ Which of the following is a location managed for current harvest and growth of a species?
   A. Fully exploited fishery
   B. Collapsed fishery
   C. Sustainable fishery
   D. Recovering fishery

38. _____ Which of the following is a location that is on an upswing, showing an increase in reproduction and growth?
   A. Recovering fishery
   B. Fully exploited fishery
   C. Collapsed fishery
   D. Overexploited fishery

39. _____ Which of the following is a location that has been overfished, reducing the potential for future reproduction and growth?
   A. Collapsed fishery
   B. Recovering fishery
   C. Sustainable fishery
   D. Overexploited fishery

Name ______________________________

40. _____ Which of the following is a location that is considered lost with little to no chance for reproduction or growth?
   A. Sustainable fishery
   B. Fully exploited fishery
   C. Overexploited fishery
   D. Collapsed fishery

41. _____ *True or False?* Catch per unit effort is a direct measure of the size of an aquatic species' population.

For questions 42–47, match the freshwater fish to the correct name.

42. _____ Black crappie (*Pomoxis nigromaculatus*)

43. _____ Channel catfish (*Ictalurus punctatus*)

44. _____ Lake trout (*Salvelinus namaycush*)

45. _____ Largemouth bass (*Micropterus salmoides*)

46. _____ Smallmouth bass (*Micropterus dolomieu*)

47. _____ Walleye (*Sander vitreus*)

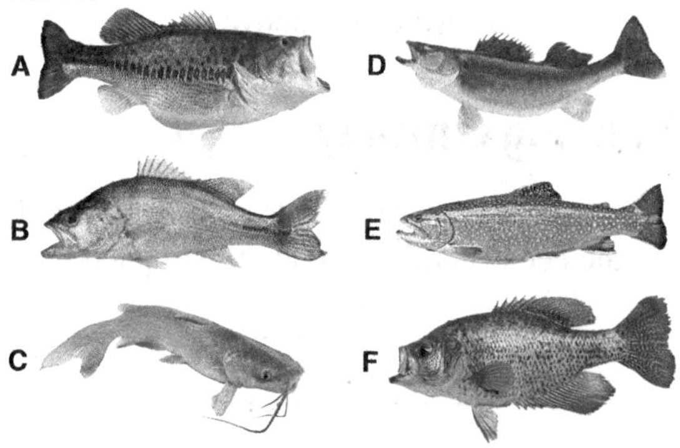

A. Steven R. Smith/Shutterstock.com, B. Alexander Sviridov/Shutterstock.com, C. Jennifer White Maxwell/Shutterstock.com, D.–F. Dan Thornberg/Shutterstock.com

For questions 48–55, match the commercial marine finfish to the correct name.

48. _____ Alaska Pollock (*Gadus chaalcogrammus*)

49. _____ Atlantic cod (*Gadus morhua*)

50. _____ Atlantic herring (*Clupea harengus*)

51. _____ Flatfish (Pleuronectiform order)

52. _____ Menhaden (*Brevoortia tyrannus*)

53. _____ Pacific skipjack tuna (*Katsuwonus pelamis*)

54. _____ Salmon (Salmoniformes order)

55. _____ Silver hake (*Merluccius bilinearis*)

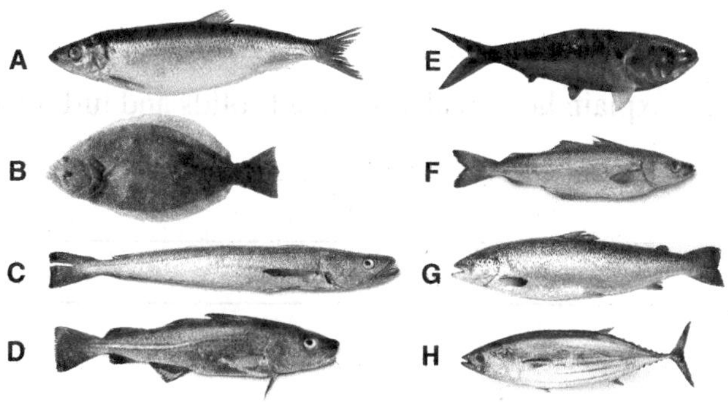

A. BigTunaOnline/Shutterstock.com, B. IrinaK/Shutterstock.com, C. Edward Westmacott/Shutterstock.com, D. picturepartners/Shutterstock.com, E. Abeselom Zerit/Shutterstock.com, F. picturepartners/Shutterstock.com, G. Edward Westmacott/Shutterstock.com, H. funny face/Shutterstock.com

56. _____ Which of the following are the leading marine species landed by US commercial fishers?
   A. Scallops, oysters, and mussels
   B. Shrimp, crabs, and squid
   C. Clams, crayfish, and lobsters
   D. All of the above.

57. _____ Which of the following is true about shrimp landings?

A. They are sustainably managed and responsibly harvested

B. Shrimp trawling often rates highest in bycatch

C. They are often referred to as an annual crop

D. All of the above.

58. _____ Which crabs are harvested on the East Coast?

A. Stone crab, blue crab, and Jonah crab

B. Red king crab, blue king crab, and golden crab

C. Tanner crab, stone crab, and Dungeness crab

D. All of the above.

Thinking Critically

1. Water erosion of soil also has negative effects on other parts of the ecosystem. What are these parts of the ecosystem? What are the impacts of water erosion in these areas?

____

____

____

____

____

____

2. Explain how total suspended solids and turbidity affect the health of a waterway.

____

____

____

____

____

____

3. Which conservation practices can farmers use on their farms to improve water quality in nearby rivers and streams? Explain how two of these practices maintain or improve water quality.

____

____

____

____

____

____

Name ______________________ Date ______________ Period __________

Activity 20A

Environmental Science: Tragedy of the Commons (Part 1)

Objectives

- Harvest as many fish as you can without destroying the lake.
- Understand the difficulty of managing shared resources.
- Understand the concept of the tragedy of the commons.
- Understand the conditions that lead to the tragedy of the commons.

Materials

- activity sheet
- fish (multicolored goldfish crackers)
- money (from board game)
- plastic bowls (one for each group)
- plastic spoons (one per student)
- chopsticks (one pair per student)
- timers (one per group)

Shared Resources

In this activity, you will work as fishers to earn a living. There is one lake that serves as the only food source. The fish in the lake represent the resources available to everyone in the group. The carrying capacity of the lake is 16 fish and each fish is worth a certain amount of money. Can you predict how the members of your group will behave as they compete for the resources?

Game Preparation

1. Divide the class into groups of four and have each group sit around a lake (bowl).
2. Each fisher in the group is the only source of income for his or her family and must sell his or her daily catch to earn money. Each fisher must catch enough fish to pay his or her operating costs and make a profit for living expenses.
3. Distribute $30 in play money to each student and place 16 fish in each bowl.
4. Have each person purchase their fishing gear. The fishing pole (chopsticks) has a one-time fee of $1, and the fishing net (spoon) has an annual fee of $3.

Go Fish!

The rules for the first round are no talking and no bumping or pushing other fishers. You will be charged a fine of $5 if you break either rule.

1. Each fisher will get a chance to fish once a year (which lasts 45 seconds) and each time you fish, you may take 1, 2, 3, or 4 fish from the lake. However, you are only allowed to catch one fish at a time. Remember, you must catch enough fish to cover your yearly expenses.
2. You catch the fish by plucking them from the lake with your chopsticks (fishing pole), or scooping them with a spoon (net). Set a timer to ensure nobody's fishing season lasts more than 45 seconds.
3. Keep your catch in front of you until everyone has taken their turn fishing.
4. Write the number of fish caught on your worksheet. If you are using only gold-colored fish, each fish is worth $10. If you are using multiple colors of fish, each one has a different value.

   Gold fish = $6 each

   Green fish = $7 each

   Red fish = $9 each

   Blue fish = $13 each
5. The fish remaining in the lake after the annual fishing season has ended, will reproduce once. Each fish produces one new fish with a maximum total of 16 fish in the lake. The multicolored fish produce the same color offspring. Your instructor will add one new fish for each remaining fish after each round (year).
6. If your group has fish remaining in the lake, run another fishing season. Once the fish are exhausted, total your worksheets.
7. Fishers who are unable to catch enough fish to support their fishing operation or run out of money, go bankrupt and are unable to fish the next year. They lose their equipment and are out of the game.

Second and Third Seasons

In these rounds, a free exchange of information is encouraged, but there is no bumping or pushing. Repeat the fishing game and record the information. Rotate your fishing order each year so that everyone has a chance to go first. Continue until all fish are exhausted or the predetermined amount of time is met. A bonus will be given to the student in each group who has accumulated the most wealth at the end of the entire simulation

Name ____

Budget Sheet: Year 1

Annual Budget		
Year 1 Operating Costs		
	Opening Balance	$30.00
Annual Fees		
	Fishing permit	$5.00
	Boat maintenance/fuel/slip fees	$5.00
	Living expenses	$10.00
	Equipment rental (net)	$3.00
Additional Equipment		
	Fishing pole	$1.00
Fines		
		$
	Total operating costs (year one)	$
Year 1 Income/Profits		
Gold fish (only type of fish available)		
	________ at $10.00 each	$
Multicolored fish		
	________ Gold fish at $6.00 each	
	________ Green fish at $7.00 each	
	________ Red fish at $9.00 each	
	________ Blue fish at $13.00 each	
	Total income/profits	$
	Total profits	$
	Operating costs (subtract)	$
	Year 1 net profit	$

Goodheart-Willcox Publisher

Budget Sheet: Year 2

Annual Budget		
Year 2 Operating Costs		
	Opening Balance	$
Annual Fees		
	Fishing permit	$5.00
	Boat maintenance/fuel/slip fees	$5.00
	Living expenses	$10.00
	Equipment rental (net)	$3.00
Additional Equipment		
	Fishing pole	—
Fines		
		$
	Total operating costs (year two)	$
Year 2 Income/Profits		
Gold fish (only type of fish available)		
	________ at $10.00 each	$
Multicolored fish		
	________ Gold fish at $6.00 each	
	________ Green fish at $7.00 each	
	________ Red fish at $9.00 each	
	________ Blue fish at $13.00 each	
	Total income/profits	$
	Total profits	$
	Operating costs (subtract)	$
	Year 2 net profit	$

Goodheart-Willcox Publisher

Name ______________________________

Budget Sheet: Year 3

Annual Budget		
Year 3 Operating Costs		
	Opening Balance	$
Annual Fees		
	Fishing permit	$5.00
	Boat maintenance/fuel/slip fees	$5.00
	Living expenses	$10.00
	Equipment rental (net)	$3.00
Additional Equipment		
	Fishing pole	—
Fines		
		$
	Total operating costs (year three)	$
Year 3 Income/Profits		
Gold fish (only type of fish available)		
	________ at $10.00 each	$
Multicolored fish		
	________ Gold fish at $6.00 each	
	________ Green fish at $7.00 each	
	________ Red fish at $9.00 each	
	________ Blue fish at $13.00 each	
	Total income/profits	$
	Total profits	$
	Operating costs (subtract)	$
	Year 3 net profit	$

Goodheart-Willcox Publisher

Notes

Name ______________________ Date ______________ Period __________

Activity 20A

Environmental Science: Tragedy of the Commons (Part 2)

Final Profit		
Year 1 Total profit	$	
Year 2 Total profit	$	
Year 3 Total profit	$	
	Total profit	$
Year 1 Operating costs	$	
Year 2 Operating costs	$	
Year 3 Operating costs	$	
	Subtract total operating costs	$
	Total net profit	$

Goodheart-Willcox Publisher

1. Did you make a profit? Did you go bankrupt? Explain what happened.

____

____

____

____

Mine, Yours, and Ours?

Originally an economic principle first explained in 1833 by William Forster, the Tragedy of the Commons explains that people will act in their own best interest when a shared resource is available and not aim to help the greater good. In 1833 Britain, shared resources were known as "commons" and Forster believed that if multiple people were given access to the same common resource, each would be selfish and consume as much as they could, in the end depleting the common resource completely.

The tragedy of commons occurs when individuals exploit a shared resource to the extent that demand overwhelms supply and the resource becomes unavailable to some or all the individuals. Ecologist Garret Hardin's theory states that individuals acting rationally and independently according to their own self-interest will deplete a shared resource, even if it is contrary to the best interest of the group. The state of fisheries and commercial fishing is a good example of Hardin's theory, as so many of our fisheries have been overfished.

- Fishing provides an income and each fisher has his or her own best interest in mind.
- Each fisher will try to catch as many fish as possible, even if all the others are also trying to catch as many fish as possible.
- The shared resource is eventually depleted.
- This behavior continues and, in the case of fisheries, the global population suffers.
- The fishers have two choices, to continue overfishing and face the consequences, or to restrict each fisher's catch limit and allow the populations to breed and grow.

Post Lab Discussion

1. Did anyone in your group take too many fish? How did you feel when somebody took too many fish?
2. Did everyone try to take as many as possible? Why or why not?
3. Did your group discuss your actions and strategies before each harvest during the second and/or third round? If so, briefly explain the discussion. Did each member carry out the plan that was discussed?
4. Did people behave or strategize differently when they knew what the rewards were? Explain why or how people behaved differently.
5. You received money for your fish harvest. Is it better to have money than fish? What can you do with money that you cannot do with fish?
6. Is it possible to maximize the number of fish caught by each person and the number of fish remaining in the pond at the same time? Why or why not?
7. Did a particular species of fish disappear faster than others? How does this relate to economically valuable species in nature and their extinction rates?
8. What are some everyday commons that people share, such as parking lots, bathrooms, and playground equipment? Do similar situations arise? Explain your answer. How might these problems be solved?
9. What are the global commons? Are these being used wisely? Why or why not?
10. What can people do to use these resources more wisely?

Name ______________________________ Date ______________ Period __________

CHAPTER 21 Game Species

Chapter Review

Carefully study the chapter and then answer the following questions.

1. _____ According to the USFWS, what amount of money is spent on fishing and hunting activities annually?
   A. $100 million
   B. $125 million
   C. $150 billion
   D. $175 billion
2. _____ *True or False?* Mammals or birds that are hunted for food or sport are game species.
3. _____ *True or False?* Hunting is used solely for subsistence.
4. _____ The grouping or organizing of animals into an outline structure is _____.
   A. specific epithet nomenclature
   B. binomial nomenclature
   C. taxonomy
   D. All of the above.
5. _____ *True or False?* White-tailed deer are the least common ungulate in North America.
6. _____ Hunters wear blaze orange clothing to ensure they are visible to other hunters. Why can hunters wear blaze orange and *not* call undue attention from the white-tailed deer?
   A. Most deer only see in black and white
   B. The deer have trouble distinguishing orange
   C. Some deer only see in blue and red
   D. All of the above.

For questions 7–14, match the common name of each game species with its scientific name.

A. bighorn sheep
B. Canada moose
C. caribou
D. elk
E. mule deer
F. pronghorn
G. white-tailed deer
H. wild boar

7. _____ *Alces alces*

8. _____ *Sus scrofa*

9. _____ *Antilocapra americana*

10. _____ *Ovis canadensis*

11. _____ *Cervus canadensis*

12. _____ *Odocoileus virginianus*

13. _____ *Rangifer tarandus*

14. _____ *Odocoileus hemionus*

15. _____ *True or False?* Most natural predators of white-tailed deer have been extirpated from eastern North America.

16. _____ Which of the following is *not* true about mule deer?
    A. They can leap as far as 8 yards and run up to 45 mph
    B. They have ears similar to mules' ears
    C. They are the smaller cousin of white-tailed deer
    D. The males shed their antlers annually

17. _____ Which of the following are often referred to as reindeer?
    A. Caribou
    B. Mule deer
    C. Elk
    D. White-tailed deer

18. _____ *True or False?* Caribou are ruminants that feed on grasses, lichens, and moss.

19. _____ *True or False?* Female caribou (cows) may grow antlers.

20. _____ Which of the following is the largest of the deer family?
    A. Caribou
    B. Pronghorn
    C. White-tailed deer
    D. Canada moose

21. _____ Which of the following live in gangs?
    A. Pronghorn
    B. Elk
    C. Caribou
    D. White-tailed deer

Name ______________________________

22. _____ Which of the following do *not* form herds?
   A. White-tailed deer
   B. Mule deer
   C. Canada moose
   D. All of the above.

23. _____ Which of the following statements are true about pronghorn?
   A. Its Latin name means American goat-antelope
   B. They can run more than 53 mph over short distances
   C. They are indigenous to western and central North America
   D. All of the above.

24. _____ Which of the following is an invasive species in North America?
   A. *Rangifer tarandus*
   B. *Ovis canadensis*
   C. *Sus scrofa*
   D. *Cervus canadensis*

25. _____ *True or False?* Bighorn sheep rams live with their flocks for the entire year.

For questions 26–29, match the game birds to the appropriate characteristics and terms.

A. Bobwhite quail
B. Canada goose
C. Ring-necked pheasant
D. Wild turkey

26. _____ *Branta canadensis*; native to northern Canada; wingspan up to 6′

27. _____ *Colinus virginianus*; ground-dwelling bird; listed as near threatened

28. _____ *Meleagris gallopavo*; gobbler, hen, jake, poults; largest North American game bird

29. _____ *Phasianus colchicus*; roosters, hens, chicks; native to Asia

For questions 30–36, match the firearms and bows to corresponding description.

A. compound bow
B. crossbow
C. longbow
D. muzzleloader
E. recurve bow
F. rifle
G. shotgun

30. _____ Fires one projectile through a grooved barrel

31. _____ More difficult to aim and has less velocity than other bows

32. _____ Complex and powerful bow used by experienced archers

33. _____ Has the gunpowder loaded directly into the barrel

34. _____ Fires several projectiles through a short barrel

35. _____ Have short firing range and requires heavier draw weights

36. _____ owerful and requires less strength than other bows

37. _____ Which of the following actions can help ensure a hunter's safety?

A. Unloading firearms when crossing obstacles

B. Wearing safety colors, such as blaze orange

C. Placing the weapon on the ground when crossing obstacles

D. All of the above.

38. _____ Which of the following is true about zoonotic diseases?

A. They are transmitted from animals to humans

B. They are transmitted through insects

C. They can infect domesticated animals

D. All of the above.

39. _____ *True or False?* Hunters are discouraged from eating meat harvested from a deer exhibiting chronic wasting disease.

40. _____ hich of the following hunters are poachers?

A. Hunters who kill more animals than regulations allow

B. Hunters who hunt out-of-season

C. Hunters who hunt species that are protected

D. All of the above.

41. _____ *True or False?* When hunting on their own property, owners have the right to hunt any species of animal at any time of the year.

Name ____

Thinking Critically

1. Identify three actions or tools that are used for game management.

____

____

____

2. What are the four basic rules of firearm safety?

____

____

____

3. List five principles of the ethical hunter.

____

____

____

4. When can a property owner obtain a kill permit or kill game animals, such as cougars or black bears?

____

____

____

5. In some states, it is not illegal to run down coyotes or wolves with a snowmobile. Is this ethical? Why or why not? Could a person in the same states run down a cat or dog without reprimand? Explain your answer.

____

____

____

6. Is something ethical or moral just because it is legal? What if something is not ethical or moral, should it be legal? Explain your answer.

____

____

____

7. Should predator killing competitions be banned? Why or why not?

____

____

____

Notes

Name ______________________________ Date ______________ Period __________

Activity 21A

Casting Animal Tracks

Objectives

- Identify tracks made by various animals.
- Determine habitat and range of various animals.

Materials

- disposable stirring spoons or sticks
- disposable containers, plastic cups or bowls, or plastic bags
- flexible cardboard strip
- measuring cup
- paper clip
- permanent marker
- plaster of Paris or dental stone (avoid using patching cement or hobby, modeling, and molding plaster)
- plastic containers to hold casts while drying
- pocket guide on animal tracks for your area (a laminated guide is more durable) with a ruler on one side (there are also apps you can put on your phone)
- salt* (optional)
- small garden shovel
- small paper notebook
- tin cans with both ends removed (coating the inside with oil makes cast removal easier)
- twine, wire, twigs (optional reinforcing materials)
- vinegar* (optional)
- water (if not available at site)

*You can add a small amount of salt to the mixture to speed up the setting of the plaster. If you add a small amount of vinegar to the mixture, you can slow the plaster's setting. Adding vinegar or salt will modify the plaster's composition and may result in an even softer cast.

Material Preparation

Measure your materials before going outside.

- The plaster can be premeasured and placed in the mixing containers if they have lids. It may also be placed in plastic baggies.
- Water can be measured and placed in recycled water bottles. This eliminates the need for measuring while in the field and each person can carry their own supply.

Locating and Preparing Tracks

The most likely places to find animal tracks are in areas with a lot of mud or sand, such as trails through the woods, along edges of streams, ponds, or rivers, and under bridges. Areas where two habitats meet may also be a good place to search for tracks. Tracks made in mud last longer and hold their shape under pressure, which makes them easier to cast. Tracks made in clayey soil are more difficult to remove than those made in sandy soil, and are more likely to break.

1. Find a track with clean features. Look for signs of animal activity, such as feathers on the ground or gnaw marks on trees.
2. Carefully remove leaves and other debris without altering the print. Do *not* attempt to remove leaves, twigs, or stones that are imbedded in the track, as it will damage the track.
3. Depending on how you will be containing the plaster, you may want to dig or build a small wall of dirt around the print. If you are using a tin can or cardboard strip to encircle the print, you will not need to dig or build a small wall of dirt around it.
4. If you are using a cardboard strip, bend it around the track and hold it in place with a paper clip. Gently press the strip into the soil to prevent the plaster from leaking at its base. If you are using a tin can, place the can over the track and gently press it into the soil.

Mixing the Plaster

Do *not* mix the plaster until you are ready to pour it into the track because it begins to set as soon as it comes in contact with the water. Read the instructions on the packaging to ensure you have the proper ratio and you know the time it needs to set.

1. Determine the proper water to plaster ratio according to the plaster packaging.
2. Mix enough plaster to fill the animal track(s).
3. If you are using plastic bags containing premeasured plaster, pour the premeasured water into the bag and mix immediately with a stick or spoon. If you are using a plastic container, add the water first and then pour the premeasured plaster into the container. Mix immediately.
4. Stir the plaster to remove the lumps. The plaster should have a consistency similar to pancake batter.
5. Tap the container on the ground to remove air bubbles from the mixture.

Pouring the Plaster

1. Slowly pour the plaster into the can or cardboard tube. Do *not* pour directly on the imprint because it may damage it. Pour the plaster along the edges of the print and let the plaster run into it. You may also hold a spoon or other object low over the print and pour the plaster so that it runs off the spoon.
2. Fill it to a depth of at least 1″ for a strong cast. Ensure that you fill all details to get a complete casting of the track.
3. You may add reinforcing materials, such as wire, twine, or twigs, at this time. Note that dental stone is stronger than plaster of Paris and does *not* require reinforcement. In addition, less dental stone is needed to make a strong casting.
4. Smooth the top if you would like a flat surface on the back of the mold.
5. Allow the plaster to dry at least 45 minutes before removing the cast from the ground. Refer to the package instructions for drying time.
6. Thoroughly clean your mixing utensils and containers before mixing plaster for another casting. The residue interferes with the chemical process and your second casting will be chalky and fragile.

Name ____________________

Removing the Casting

1. As it dries, the plaster changes from wet and glossy to a dull matte finish. When the finish is dull, gently touch the surface to see if it is dry. Do *not* press hard on the plaster because it may crack.
2. f the casting is dry after an hour, you may remove it from the ground. You may need to dig around and below the casting to remove it. Do *not* remove the tin can or cardboard surrounding the casting and do *not* pry the casting from the ground as it may crack. Leave the casting in place for as long as possible.
3. Lift the casting and place it in a container with the can still in place. Do *not* cover the casting with plastic because it will retain moisture instead of releasing it.
4. After the plaster dries for 24 hours, you may remove the tin can.
5. Allow the plaster to dry another day or two before handling it.
6. Gently brush or wipe excess dirt from the casting. Scrubbing too hard may remove details.
7. Use the pocket guide on animal tracks to identify the animals that made the tracks.

Notes

Name ______________________________ Date ______________ Period __________

Activity 21B

Limiting Factors

Objectives

- Determine how limiting factors play a role in animal survival and behavior in an ecosystem.
- Understand how limited resources affects the carrying capacity of an ecosystem.

Materials

paper cards:

- blue labeled water, worth two points
- green labeled plants, worth one point
- orange labeled mast, worth one point
- red labeled meat, worth one point
- large area for students to move around in (gym, soccer field, etc.)

Instructions

Your teacher will scatter the cards around the ecosystem before you begin the activity. Follow your teacher's instructions and answer the following questions as you complete each round of the carrying capacity lab.

Round One

1. Spread out around the gym or field and designate a spot as your den.
2. On your teacher's order, walk (*no* running) around the environment and collect cards. You must collect at least one blue card and at least 10 points to survive the round.
3. When all the cards are collected, return to your den.
4. If you have less than 10 points and/or no water, you did not survive and are out of the game.

After Round One

1. Did you successfully collect the resources?

   ____

2. What was your strategy for collecting the resources?

   ____

   ____

   ____

3. What would you change about your performance this round?

   ____

   ____

   ____

Round Two

1. Your teacher will remove 10 random cards from play and scatter the remaining cards around the environment.
2. Not all bears are equally prepared to survive. In this round, your teacher will designate some students as having cubs. Bears with cubs must collect at least two water cards and earn at least 20 points. Other bears will also be given disadvantages; such as being blindfolded or crippled. These bears must collect at least one water card and a total of 12 points to survive. The remaining healthy bears must collect at least one blue card and a total of 10 points.
3. On your teacher's order, walk (*no* running) around the environment and collect cards.
4. When all the cards are collected, return to your den.
5. If you have less than the required points and/or no water, you did not survive and are out of the game.

After Round Two

1. Did you successfully collect the resources?

____

2. What was your strategy for collecting the resources?

____

____

____

3. What would you change about your performance this round?

____

____

____

Round Three

1. In this round, the area is experiencing a severe drought. Your teacher will remove 1/2 of the water cards and 1/2 of the food cards from play.
2. You are still the same bear from the second round if you survived. Bears with cubs must collect at least two water cards and earn at least 20 points. The remaining bears must collect at least one blue card and a total of 10 points.
3. On your teacher's order, walk (*no* running) around the environment and collect cards.
4. When all the cards are collected, return to your den.
5. If you have less than the required points and/or no water, you did not survive and are out of the game.

Name ____

After Round Three

1. Did you successfully collect the resources?

____

2. What was your strategy for collecting the resources?

____

____

____

3. What would you change about your performance this round?

____

____

____

Final Question

1. How can limiting factors play a role in animal survival and behavior in an ecosystem?

____

____

____

____

____

Notes

Name ______________________________ Date ______________ Period __________

CHAPTER 22 Forests

Chapter Review

Carefully study the chapter and then answer the following questions.

1. _____ Which of the following is a type of tree that remains green year-round and bears needles instead of leaves?
   A. Broadleaf
   B. Coniferous
   C. Deciduous
   D. Canopy

2. _____ Which type of forest is composed almost entirely of trees with flat leaves that drop during the fall season?
   A. Deciduous forest
   B. Evergreen forest
   C. Wet forest
   D. Coniferous forest

3. _____ Which of the following statements is true about needle-bearing trees?
   A. They have less surface area to lose moisture
   B. They have less surface area to hold snow and ice
   C. They perform photosynthesis year-round
   D. All of the above.

4. _____ *True or False?* Deciduous forests typically occupy high elevations on mountainsides.

5. _____ *True or False?* Mixed forests have more deciduous trees than coniferous trees.

6. _____ The uppermost layer of a mixed forest _____.
   A. is made up of tall, mature trees
   B. has trees that are shade intolerant
   C. is called the canopy
   D. All of the above.

7. _____ *True or False?* The understory of a mixed forest consists of herbaceous plants and grasses.

8. _____ *True or False?* Trees reproduce sexually by producing seeds, which are the result of fertilization and pollination of the female sex cell in the tree.

9. _____ Which of the following are fleshy fruits?

A. Legume seed pod

B. Walnuts and hickory nuts

C. Drupes, berries, pomes

D. All of the above.

10. _____ *True or False?* Indehiscent dry fruits split open along predetermined lines at maturity.

11. _____ *True or False?* Dehiscent dry fruits do *not* split open along predetermined lines at maturity.

12. _____ Seed dispersal in the forest requires the mechanical action of an organism or agent other than the tree. Which of the following are agents of seed dispersal?

A. Water and wind

B. Gravity and animals

C. Birds and squirrels

D. All of the above.

13. _____ Plants use carbon dioxide to produce sugars to use during which of the following processes?

A. Respiration

B. Transpiration

C. Photosynthesis

D. Germination

For questions 14–23, match the tree species to the corresponding characteristics.

A. balsam fir
B. Douglas fir
C. flowering dogwood
D. loblolly pine
E. lodgepole pine
F. maple
G. quaking aspen
H. sugar maple
I. sweetgum
J. white oak

14. _____ *Acer rubrum*; most common tree in the United States; grow 40′ to 70′

15. _____ *Acer saccharum*; used to make cabinets, cutting boards, and baseball bats; grow 40′ to 80′

16. _____ *Cornus florida*; have white flowers in spring; used as short landscape trees; grow 15′ to 30′

17. _____ *Abies balsamea*; used for framing timbers and pulp or paper; grow 45′ to 75′

18. _____ *Liquidambar styraciflua*; grow 60′ to 75′; used for shade, lumber, veneer, and pulpwood

19. _____ *Quercus alba*; wood is used for furniture, cabinetry, baskets, flooring, wine and whiskey barrels; 50′ to 80′ tall

20. _____ *Pinus contorta*; tall and skinny, growing to 70′ to 80′; used for construction, telephone poles, railroad ties, and pulpwood

21. _____ *Pinus taeda*; native to the southeastern United States; fast growing, reaching heights of 40′ to 90′; used for lumber and pulpwood

22. _____ *Populus tremuloides*; largest range of any tree in North America; grow in stands called clonal groves and reach heights of 20′ to 50′

23. _____ *Pseudotsuga menziesii*; major contributor to construction timber; used for telephone poles, railroad ties, and half of all Christmas trees grown in United States

Name ____

24. _____ This species has nitrogen-fixing nodules on its roots, similar to legumes.

A. Green ash

B. American beech

C. Red alder

D. American basswood

25. _____ One of the largest (165′ to 180′) and most common of the pine trees in the eastern United States.

A. *Pinus taeda*

B. *Pinus strobus*

C. *Pinus contorta*

D. *Pinus menziesii*

26. _____ This species grows 50′ to 70′ and has a range from the Rocky Mountains to the Atlantic Ocean.

A. Green ash

B. White ash

C. Yellow birch

D. American basswood

27. _____ Which of the following are economically important species?

A. Hickory and fir

B. Oak and walnut

C. Maple and pine

D. All of the above.

28. _____ This slow-growing species is prized as shade trees because of their spreading canopy. They provide nuts for forest animals, such as squirrels, deer, and chipmunks.

A. *Betula papyrifera*

B. *Fagus grandifolia*

C. *Juglans cinerea*

D. *Prunus serotina*

29. _____ This slow-growing species, which rarely lives longer than 75 years, is often called white walnut.

A. Black cherry

B. Butternut

C. Cottonwood

D. River birch

30. _____ This species produces a fleshy fruit that serves as food for many forest animals. The wood is often used for cabinetry.

A. *Prunus serotina*

B. *Carya ovata*

C. *Ilex opaca*

D. *Robina pseudoacacia*

31. _____ These slow-growing trees were once abundant in river bottoms and on other low-lying land, but are now rare. The nuts are heavy and not dispersed easily.

A. Cottonwood

B. Shagbark hickory

C. Shellbark hickory

D. Black walnut

For questions 32–41, match the tree species common names to the proper scientific name.

A. American basswood

B. American holly

C. black locust

D. butternut

E. cottonwood

F. paper birch

G. river birch

H. shagbark hickory

I. southern magnolia

J. yellow birch

32. _____ *Tilia Americana*

33. _____ *Betula papyrifera*

34. _____ *Betula nigra*

35. _____ *Betula alleghaniensis*

36. _____ *Juglans cinerea*

37. _____ *Populus deltoides*

38. _____ *Carya ovata*

39. _____ *Ilex opaca*

40. _____ *Robina pseudoacacia*

41. _____ *Magnolia grandiflora*

42. _____ Fast-growing trees with strong wood and a long life span. Growing 60′ to 90′, these trees may grow 24 a year.

A. *Liriodendron tulipfera*/tulip poplar

B. *Maclura pomifera*/osage-orange

C. *Carya illinoinensis*/pecan

D. *Diospyros virginiana*/persimmon

43. _____ Small, shrublike trees, typically growing 30′ to 50′. They have strong, dense wood and dense thorns on their branches.

A. *Diospyros virginiana*/persimmon

B. *Maclura pomifera*/osage-orange

C. *Carya illinoinensis*/pecan

D. *Taxodium distichum*/bald cypress

Name ______________________________

44. _____ Long-living but short-growing trees that may spread their branches as much as 90′.

A. *Platanus occidentalis*/sycamore

B. *Diospyros virginiana*/persimmon

C. *Quercus virginiana*/live oak

D. *Maclura pomifera*/osage-orange

45. _____ Fruit-bearing trees that are dioecious and grow 30′ to 60′.

A. *Diospyros virginiana*/persimmon

B. *Carya illinoinensis*/pecan

C. *Sassafras albidum*/sassafras

D. *Taxodium distichum*/bald cypress

46. _____ Small trees or large shrubs that grow 20′ to 30′ with a twisting trunk and widely spreading branches.

A. *Carya illinoinensis*/pecan

B. *Taxodium distichum*/bald cypress

C. *Salix nigra*/black willow

D. *Cersis Canadensis*/redbud

47. _____ Tall-growing trees (70′ to 100′) that are native to the southern United States. Produce drupes and may live more than 300 years.

A. *Taxodium distichum*/bald cypress

B. *Salix nigra*/black willow

C. *Carya illinoinensis*/pecan

D. *Platanus occidentalis*/sycamore

48. _____ A small tree (30′ to 60′) with leaves that come in three different shapes.

A. *Sassafras albidum*/sassafras

B. *Diospyros virginiana*/persimmon

C. *Cersis Canadensis*/redbud

D. *Carya illinoinensis*/pecan

49. _____ Deciduous conifers that grow in southern swamps. These trees access air through knees that stick up above the water.

A. *Platanus occidentalis*/sycamore

B. *Taxodium distichum*/bald cypress

C. *Salix nigra*/black willow

D. *Diospyros virginiana*/persimmon

50. _____ Large trees that grow 75′ to 100′ with trunks as wide as 13′. They are also called American planetrees.

A. *Sassafras albidum*/sassafras

B. *Carya illinoinensis*/pecan

C. *Platanus occidentalis*/sycamore

D. *Taxodium distichum*/bald cypress

51. _____ Native to the eastern United States, these trees prefer wet, swampy soils. These trees absorb contaminants from the soil.

A. *Sassafras albidum*/sassafras

B. *Salix nigra*/black willow

C. *Platanus occidentalis*/sycamore

D. *Diospyros virginiana*/persimmon

52. _____ *True or False?* Allelopathy occurs when one plant releases toxins that prevent the growth of nearby plants.

53. _____ A small, woody-stemmed plant that makes up the understory of a forest is a _____.

A. liverwort

B. fern

C. shrub

D. sedge

For questions 54–58, match the small, woody-stemmed plants to the corresponding description.

A. buckthorn

B. hazel

C. honeysuckle

D. rhododendron

E. viburnum

54. _____ *Corylus americana*; nut-producing shrub that grows 15′ to 18′

55. _____ *Viburnum lentago*; low-growing flowering shrub that produces white flowers

56. _____ *Rhododendron maximum*; grows 5′ to 15′; produces white, pink, or purple flowers

57. _____ *Lonicera periclymenum*; arching shrubs and climbing vines that have sweet-smelling yellow or white flower in early spring

58. _____ *Rhamnus cathartica*; an invasive shrub with spiny, thorny branches; it is not poisonous to birds but is poisonous to humans and other animals

59. _____ *True or False?* Grasses often grow abundantly after a tree falls and opens a canopy.

60. _____ *True or False?* Sedges are grasses that resemble grasses, except they have round stems, whereas grasses have triangular stems.

61. _____ Woody and semi-woody plants that grow by clinging to and climbing up other structures are _____.

A. forbs

B. sedges

C. vines

D. liverworts

62. _____ Which of the following plants lack a vascular system?

A. Mosses and liverworts

B. Club mosses and ferns

C. Vines and sedges

D. All of the above.

Name ____

63. _____ Which of the following is a fungus?

A. Lichens

B. Mushrooms

C. Puffballs

D. All of the above.

64. _____ *True or False?* Lichens are considered perennial plants.

65. _____ *True or False?* Mushrooms are fungi that produce a cap containing spores.

66. _____ Wood that has been cut into boards of various sizes is referred to as _____.

A. plywood

B. veneer

C. lumber

D. All of the above.

67. _____ *True or False?* Pulpwood is used to make veneers.

68. _____ *True or False?* All trees sequester the same amount of carbon.

Thinking Critically

1. Identify three reasons why healthy forest ecosystems are indispensable.

____

____

____

____

____

____

2. What are the four main forces that trigger germination?

____

____

____

____

____

____

3. What are the five main mechanisms of vegetative reproduction in the forest?

4. Do you agree or disagree with the following statement? Favorable economics justifies all actions; if it is economical, it is all right. Explain your answer.

5. How can science help us make forestry more sustainable? What are the limitations of science to help solve problems?

6. What should we conserve when we consider conservation issues? Should we conserve human activities and traditions as well as plants and animals?

Name ______________________________ Date ______________ Period __________

Activity 22A

Reading Tree Cuttings

Objectives

- Determine the age of trees from cut pieces.
- Identify weather phenomena that affected tree growth at various times in its life.

Materials

- activity sheet
- three cross cuttings of different tree trunks (crosscut disks can be purchased at many craft stores)
- measuring tape or ruler
- pencil

Instructions

Use the sample disks to complete the following questions.

Disk One

1. How old was this tree?

   ____

2. Please note years where the annual rings are thicker or thinner. Count the center of the disk as year one.

   A. Thicker years: ______________

   B. Thinner years: ______________

3. What is the diameter of this section of the trunk (diameter of disk)?

   ____

Disk Two

1. How old was this tree?

   ____

2. Please note years where the annual rings are thicker or thinner. Count the center of the disk as year one.

   A. Thicker years: ______________

   B. Thinner years: ______________

3. What is the diameter of this section of the trunk (diameter of disk)?

   ____

Disk Three

1. How old was this tree?

___

2. Please note years where the annual rings are thicker or thinner. Count the center of the disk as year one.

   A. Thicker years: ___

   B. Thinner years: ___

3. What is the diameter of this section of the trunk (diameter of disk)?

___

Conclusion

1. Why do you think there are differences in the thickness of the annual rings?

___

___

___

___

___

___

2. What information can trees give us about past climates?

___

___

___

___

___

___

___

On Your Own

1. Obtain a crosscut section of a tree recently cut down in your area.
2. Answer the same questions you answered in regards to the disks in class.
3. Obtain weather reports (including rainfall, temperature, and natural phenomena such as tornadoes and hurricanes) covering the life span of the tree.
4. Compare the variances of the rings (i.e., width and color) to the weather reports. Note how the rings vary in times of drought, longer winters, forest fires, etc.
5. Are the answers in your conclusion section of the activity accurate?

Name ______________________________ Date ______________ Period ____________

Activity 22B

Calculating Board Feet

Objective

- Calculate board feet with provided information.

Materials

- activity sheet
- calculator

Instructions

Use the following calculations to determine the board feet (measurement of the volume of wood) of the boards listed in the chart. Convert the fractions to decimals before performing your calculations.

Width in inches × length in feet × thickness in

inches ______________________________ 12

-or-

Width in inches × length in feet × thickness in

inches ______________________________ 144

Board #	Thickness	Length	Width	Board Feet
1	1S″	8′	3S″	
2	3S″	8′	3S″	
3	1S″	16′	5S″	
4	1″	4′	8″	
5	1S″	84″	3S″	
6	2″	24″	4″	
7	3S″	18″	3S″	
8	1″	12″	1″	
9	1″	12′	1″	
10	1S″	192″	3S″	

1. Why is the ability to calculate board feet important to someone working in the forestry industry?

____

____

____

____

Notes

Name ______________________________ Date ______________ Period __________

CHAPTER 23 Forest Succession and Management

Chapter Review

Carefully study the chapter and then answer the following questions.

1. _____ *True or False?* Foresters use natural succession as part of their management program or to combat the forces that cause succession.

2. _____ The natural evolution and changes of a forest is the process of _____.
   - A. decomposition
   - B. elimination
   - C. succession
   - D. delineation

3. _____ The colonization of an area that previously had no living ecology is _____.
   - A. secondary succession
   - B. primary succession
   - C. tertiary succession
   - D. secondary colonization

4. _____ Which of the following is *not* a characteristic of a pioneer species?
   - A. They produce abundant seeds that disperse readily
   - B. They grow quickly and mature early
   - C. They are almost exclusively shade-tolerant species
   - D. Their seeds germinate quickly

5. _____ Which of the following occurs where there is a severe disturbance in a preexisting ecological community?
   - A. Primary succession
   - B. Tertiary succession
   - C. Secondary colonization
   - D. Secondary succession

6. _____ Which type of succession occurs when a forest burns completely and the soil still contains microbes and plant seeds?
   - A. Secondary succession
   - B. Tertiary succession
   - C. Intermediate succession
   - D. Primary succession

7. _____ Which of the following is a mature ecosystem?

A. Seral community
B. Climax community
C. Tertiary community
D. Secondary community

8. _____ Each of the intermediary steps in the succession process is known as a _____.

A. Tertiary community
B. Climax community
C. Seral community
D. Secondary community

9. _____ Which plants take over a pond at the climax community step?

A. Shrubs
B. Trees
C. Moss
D. Lichen

10. _____ *True or False?* Sustainable forest management balances economic, ecological, and social impacts now and in the future.

11. _____ Which of the following have input on the development of a forest management plan?

A. Citizens
B. Stakeholders
C. Businesses
D. All of the above.

12. _____ *True or False?* Sustainable forestry practices maintain biodiversity.

13. _____ *True or False?* Regeneration capacity is the amount of time it takes foresters to replant trees in an area that has been harvested.

14. _____ *True or False?* Self-pruning trees naturally shed their lower branches

15. _____ Which of the following statements about timber stand improvement is true?

A. Desirable trees may be removed for thinning purposes
B. May be used to promote wildlife habitat
C. Commercially valuable trees may be removed
D. All of the above.

16. _____ *True or False?* Shade-tolerant species are considered "weed" species because of their fast growth and low wood quality.

17. _____ Why is girdling performed?

A. To help a tree recover quickly from disease
B. It improves the quality of the wood
C. It kills the tree and allows it to decompose naturally
D. All of the above.

Name ______________________________

18. _____ Why are the lower branches removed from timber trees?
   A. They make it difficult to harvest the tree
   B. Branches create knots in the wood
   C. To improve wind resistance
   D. All of the above.

19. _____ A tree that has desirable characteristics and has the potential to outcompete nearby trees is a _____.
   A. crop tree
   B. weed tree
   C. group tree
   D. buffer tree

20. _____ Groupings of similar topography or subpopulations of species make up individual forest _____.
   A. harvests
   B. selections
   C. inventories
   D. All of the above.

21. _____ The diameter of a tree at 4.5′ (1.3 m) off the ground is the diameter at _____.
   A. knee height
   B. breast height
   C. shoulder height
   D. waist height

22. _____ *True or False?* The harvestable height of a tree is the part of the trunk without large branches.

23. _____ A board foot of timber is a unit that is _____.
   A. 2′ long, 2′ wide, and 2″ deep
   B. 1′ long, 1′ wide, and 10″ deep
   C. 1′ long, 1′ wide, and 1″ deep
   D. 2′ long, 2′ wide, and 4″ deep

24. _____ *True or False?* Only clear-cut harvesting is destructive to the forest ecology.

25. _____ Seed trees that remain after harvest are selected based on their _____.
   A. growth rate and disease resistance
   B. desirability and future marketability
   C. wind resistance and seed-producing ability
   D. All of the above.

26. _____ Shelterwood harvest is the process of removing one-third to one-half of the mature trees in an area every _____.
   A. 1 to 5 years
   B. 10 to 15 years
   C. 15 to 20 years
   D. 20 to 25 years

27. _____ With this method of harvesting, trees are harvested in a long, narrow band of land.
   A. Strip-cutting
   B. Group selection cutting
   C. Coppicing
   D. Clear-cutting

28. _____ With this method of harvesting, mature trees are harvested and immature trees are thinned in an area.
   A. Strip-cutting
   B. Group selection cutting
   C. Coppicing
   D. Clear-cutting

29. _____ *True or False?* With single tree cutting, one old, diseased, or valuable tree is removed from an area.

30. _____ *True or False?* Coppicing is a very destructive form of tree harvesting.

31. _____ *True or False?* Reforestation occurs naturally or through planting seeds or seedlings.

32. _____ *True or False?* Woodlots are used for large-scale production of veneer trees.

Vocabulary Review

Match the forestry terms to the corresponding description or definition.

A. clear-cutting
B. crop tree release
C. culling
D. double girdle
E. forest inventory
F. girdling
G. pruning
H. silviculture
I. thinning
J. timber
K. timber cruising
L. timber stand improvement

1. _____ Trees harvested for lumber, pulp, or other forest products.
2. _____ The management of forests for overall growth, species composition, tree health, and timber quality.
3. _____ A collection of practices used to improve the composition, health, and growth of forests and to maintain their productivity.
4. _____ The overall reduction in the number of trees in an area of the forest.
5. _____ A double cut through the bark, phloem, and cambium around the circumference of the tree.
6. _____ A special type of thinning in which a tree of the desired tree species is freed from competition with nearby trees.
7. _____ The practice of walking a forest to determine best timber management practices for the area.
8. _____ A collection of data that includes the species, volumes, counts, and conditions of trees in the forest.
9. _____ Cutting the tree through the bark, phloem, and cambium around the circumference of the tree.
10. _____ The removal of dead, diseased, and undesirable trees.
11. _____ The practice of removing all trees from a given area.

Name ______________________________

Thinking Critically

1. Describe the basic steps of primary succession.

____

____

2. Briefly explain pond succession.

____

____

____

3. What information does a forester gather when performing forest inventories? For what purpose is the data used?

____

____

____

4. What are the benefits of slash? Why is slash detrimental to the health of the ecosystem?

____

____

____

5. Why are prescribed burns used in forests?

____

____

____

6. If things are in conflict, what or who should get priority? For example, if citizens want to keep a natural area in the center of their city, but the city government wants to develop the property to generate income, what or who should get priority?

____

____

____

7. How can planting all the same type of tree in an area be detrimental to the ecosystem?

____

____

8. Should the type of harvesting a forester uses be regulated by the government? Explain your answer.

____

____

____

Notes

Name ______________________ Date ______________ Period __________

Activity 23A

Flower Dissection and Analysis

Objectives

- Identify the parts of a flower.
- Distinguish between male and female flowers.
- Determine if the male or female parts have an advantage in fertilization.
- Discuss how pollinators and plants work together.

Materials

- craft knife or scalpel
- drawing paper
- markers or colored pencils
- one type of flower per group
- ruler

Instructions

In this activity, you and your partner will identify the flower by its botanical and common names, draw the flower as a whole, and then carefully dissect the flower to view its parts. You will also draw the flower as it appears inside.

Procedure

1. Draw the flower. Label the visible parts, such as the sepals and petals, and use markers or colored pencils to color your drawing.
2. *Carefully* make a vertical incision to open the flower and expose the internal parts.
3. Draw and color the internal parts of your flower. Identify and label the parts (sepals, anther, stamen, filament, stigma, style, ovary, and pistil).
4. Measure the length of the pistil (stigma, style, and ovary) in millimeters.

   Length of pistil: ______________________
5. How many filaments are there? Measure the length of the filament in millimeters.

   A. Filaments: ______________________

   B. Length of filament: ______________________
6. Measure the length of the anther in millimeters. How many anthers are there?

   A. Anthers: ______________________

   B. Length of anther: ______________________
7. Are the anthers releasing pollen? Look for a powdery residue.
8. Look inside the ovary and identify ovules. When fertilized, these will become seeds.
9. The measurements from all the flowers will be charted as explained by your instructor.

Analysis

1. Name the female parts of your flower.

____

____

2. Name the male parts of your flower.

____

____

3. In your flower, which was longer? ___________________________

4. According to the class data table, which flower type had the longer lengths for the following parts?

   A. Filaments: ____________________

   B. Anthers: ____________________

   C. Pistils: ____________________

5. Was your flower mature (open or closed)? ___________________________

6. Was your flower releasing pollen? How could you tell?

____

____

____

7. Why would having a longer pistil or stamen be an advantage in pollination?

____

____

____

Conclusion

Briefly explain what you learned about flower parts and pollination.

____

____

____

____

____

____

____

____

____

____

____

Name ________________ Date ________ Period ________

Activity 23B

Forest in a Jar

Objectives

- Investigate changes in ecological systems over time.
- Determine factors which impact the growth and development of forests.
- Distinguish between biotic and abiotic factors in plant growth.

Time

30 to 40 days

Materials

- 1 or 2 1-liter clear soda bottles (one per student/small group)
- water
- soil
- aquatic plants (one per bottle)
- 1 package mixed wildflower and/or sunflower seeds (about 60 total seeds); vegetable seeds (i.e., corn, beans, or cucumbers) may also be used

Procedure

1. Using scissors, remove the top of the container so that a hand can fit inside.
2. Place 2 to 3 cups of soil and 3 to 4 cups of water into the container. Allow the container to sit overnight to settle.
3. Plant an aquatic plant in the bottle. Keep the bottle in a sunny window or use a grow light to provide the light needed for the plant to photosynthesize.
4. Do *not* replace the water that evaporates from the jar.
5. Every 3 days, add 3 or 4 seeds to the bottle. This begins while there is still standing water in the container. Observe what happens to the seeds daily. Continue adding seeds even after the water evaporates.
6. As the water level continues to fall, record what is happening with the aquatic plant and keep adding seeds, every 3 days. Once all water has evaporated, add water, 1/4 cup every 3 days or mist daily, to keep the soil damp but not wet.

Thought Questions – Answer on Day 1

1. What will happen to the forest jug environment as time passes?
2. What will happen to the aquatic plant?
3. What purpose do the seeds in the forest jar play?
4. What external factors are at work with the forest jar?

Final Questions – Answer after Experiment is Finished

1. What happened to the forest jug environment as time passed?
2. What happened to the aquatic plant?
3. What purpose did the seeds in the forest jar play for the plant? For the environment?
4. What external factors impacted the jar?
5. How does this experiment mirror wildlife plants in a natural environment?

Name ______________________ Date ______________ Period __________

CHAPTER 24 Threats to Forests

Chapter Review

Carefully study the chapter and then answer the following questions.

1. _____ *True or False?* An organism that is infectious and transmissible is a pathogenic organism.

2. _____ Which of the following are pathogens?
   A. Parasites and bacteria
   B. Fungi and viruses
   C. Nematodes
   D. All of the above.

3. _____ *True or False?* All bacteria, fungi, and viruses are detrimental.

4. _____ Which of the following statements is true about foliar disease?
   A. They require two hosts to survive and spread
   B. They cannot be spread by insects
   C. Foliar diseases affect photosynthesis and the tree's growth
   D. All of the above.

5. _____ *True or False?* Healthy, mature trees may die within weeks or a few months after being attacked by a vascular disease.

For questions 6–13, match the diseases to the corresponding description.

A. anthracnose
B. brown spot needle blight
C. leaf blister
D. needle cast disease
E. pitch canker
F. sudden oak death
G. wetwood
H. wood decay

6. _____ *Mycosphaeralla dearnessii*, causes the needles of pine trees to turn brown and die

7. _____ Fungi are the primary vectors, causes needles to turn yellow and then brown

8. _____ Caused by *Taphrina caerulescens*, infects the early stages of leaf development and causes the leaves to curl, twist, and eventually die

9. _____ A group of tree diseases that infect valuable broadleaf trees, severe outbreaks can fully defoliate a tree

10. _____ Caused by *Fusarium subglutinans*, damaging to all southern pines, spreads during windy, wet periods in summer

11. _____ A bacterial disease that infects elms, oaks, tulip poplars, and maples; causes large, dark vertical streaks in the trunk

12. _____ Caused by fungus *Phytophthora ramorum*, infects rhododendrons, Douglas firs, oaks, and redwoods

13. _____ Caused by many different fungi, including red belt fungus, chicken of the woods, and quinine conk

14. _____ Where is evidence of root disease often exhibited on a tree?

A. On the highest branches of new growth

B. In the shoots

C. In rapid-growth parts

D. All of the above.

15. _____ Which of the following fungi cause Dutch elm disease?

A. *Phytophthora cinnamomi* and *Heterobasidion annosum*

B. *Ophiostoma ulmi* and *Ophiostoma novo-ulmi*

C. *Raffaelea lauricola* and *Xylerborus glabratus*

D. *Bretziella fagacearum* and *Xylella fastidiosa*

16. _____ This disease is spread by the redbay ambrosia beetle (*Xylerborus glabratus*).

A. Oak wilt

B. Laurel wilt

C. Scorch wilt

D. Pine wilt

17. _____ The fungus that causes this disease creates a sweet-smelling mat that attracts nitidulid, or picnic beetles.

A. Oak wilt

B. Laurel wilt

C. Scorch wilt

D. Pine wilt

18. _____ Which of the following is a disease that turns the leaf margins of shade trees brown?

A. Laurel leaf scorch

B. Pitch canker

C. Bacterial leaf scorch

D. Little leaf disease

19. _____ *True or False?* Oak decline is a complex of diseases that commonly affects young oak trees under stress.

Name ____

For questions 20–27, match the forest pests to the proper names.

A

B

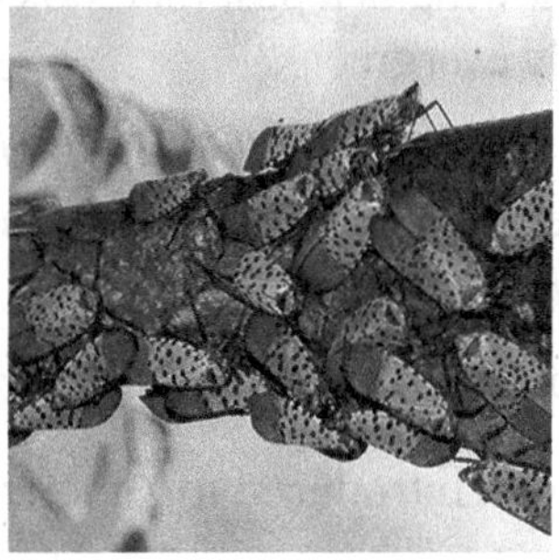

C

D

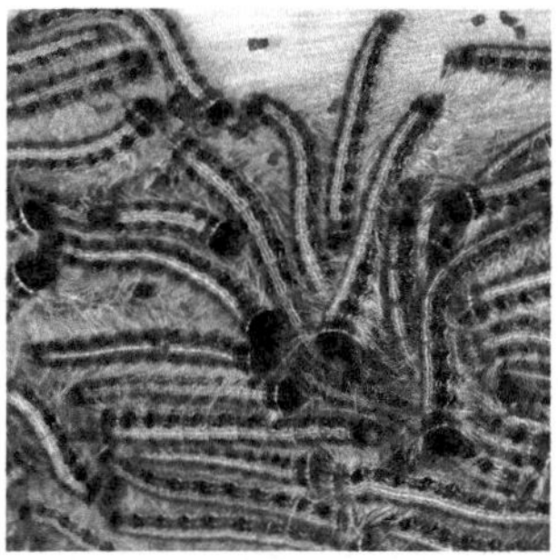

E

F

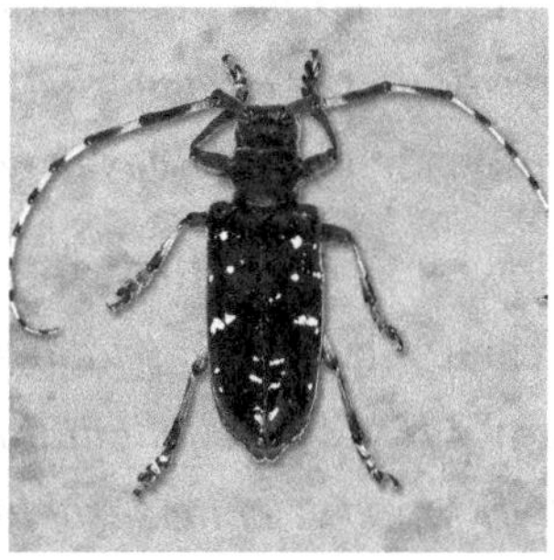

G

H

A. Herman Wong HM/Shutterstock.com; B. Tom Coleman, USDA Forest Service, Bugwood.org UGA5424552; C. Jana Shea/Shutterstock.com; D. USDA Forest Service–Region 8–Southern, USDA Forest Service, Bugwood.org UGA1510071; E. Lacy L. Hyche, Auburn University, Bugwood.org UGA1430006; F. Scott Baur, USDA Agricultural Research Service, Bugwood.org UGA1321093; G. Protasov AN/Shutterstock.com; H. Clemson University–USDA Cooperative Extension Slide Series, Bugwood.org UGA1435156

20. _____ Asian long-horned beetle/*Anoplophora glabripennis*

21. _____ eastern tent caterpillar/*Malacosoma americanum*

22. _____ emerald ash borer/*Agrilus planipennis*

23. _____ gypsy moth caterpillar/*Lymantria dispar*

24. _____ hemlock woolly adelgid/*Adelges tsugae*

25. _____ southern pine beetle/*Dendroctonus frontalis*

26. _____ spotted lanternfly/*Lycorma delicatula*

27. _____ twig girdler/*Oncideres cingulate*

28. _____ *True or False?* Defoliators are insects that feed on the leaves and needles of trees.

29. _____ Which of the following is a pest that looks like a pine cone and feeds on over 50 different tree species, primarily evergreens?

A. Gypsy moth caterpillar/*Lymantria dispar*
B. Spotted lanternfly/*Lycorma delicatula*
C. Bagworm/*Thyridopteryx ephemeraeformis*
D. Hemlock woolly adelgid/*Adelges tsugae*

30. _____ This forest pest was introduced to America in the late 1860s in Boston, Massachusetts, from Europe.

A. Gypsy moth caterpillar/*Lymantria dispar*
B. Spotted lanternfly/*Lycorma delicatula*
C. Bagworm/*Thyridopteryx ephemeraeformis*
D. Hemlock woolly adelgid/*Adelges tsugae*

31. _____ Which of the following is controlled biologically by the black lady beetle?

A. Gypsy moth caterpillar/*Lymantria dispar*
B. Spotted lanternfly/*Lycorma delicatula*
C. Bagworm/*Thyridopteryx ephemeraeformis*
D. Hemlock woolly adelgid/*Adelges tsugae*

32. _____ While feeding, this pest creates a sticky substance called honeydew that attracts other insects.

A. Gypsy moth caterpillar/*Lymantria dispar*
B. Spotted lanternfly/*Lycorma delicatula*
C. Bagworm/*Thyridopteryx ephemeraeformis*
D. Hemlock woolly adelgid/*Adelges tsugae*

33. _____ *True or False?* Bark borers kill trees by boring into the bark and down to the roots.

34. _____ This pest is the most destructive pest in the south, and is expanding north because of the warming climate.

A. Gypsy moth caterpillar/*Lymantria dispar*
B. Western bark beetles/*Dendroctonus brevicomis*
C. Southern pine beetles/*Dendroctonus frontalis*
D. Emerald ash borers/*Agrilus planipennis*

35. _____ *True or False?* Emerald ash borers have killed tens of millions of ash trees in North America and threatens to kill nearly all ash trees in North America.

36. _____ These pests drill holes in the bark, feed on the cambium layer, and then burrow farther to feed on the heartwood.

A. Pine sawyer/*Monochamus scutellatus*
B. Western bark beetles/*Dendroctonus brevicomis*
C. Southern pine beetles/*Dendroctonus frontalis*
D. Asian long-horned beetles/*Anoplophora glabripennis*

Name ____

37. _____ A woody shrub that is more obnoxious than kudzu and covers more than 3.2 million acres of forested land in the United States.

A. Chinese privet/*Ligustrum sinense*

B. Chinese tallow/*Triadica sebifera*

C. Purple loosestrife/*Lythrum salicaria*

D. Cogon grass/*Imperata cylindrica*

38. _____ Which of the following was introduced by the USDA in 1876?

A. Chinese privet/*Ligustrum sinense*

B. Chinese tallow/*Triadica sebifera*

C. Kudzu/*Pueraria montana*

D. Cogon grass/*Imperata cylindrica*

39. _____ The leaves and flowers of this species produce a toxin that make them distasteful to insects, birds, and other herbivores.

A. Chinese privet/*Ligustrum sinense*

B. Chinese tallow/*Triadica sebifera*

C. Kudzu/*Pueraria montana*

D. Cogon grass/*Imperata cylindrica*

40. _____ This invasive plant is a perennial that grows from 2′ (0.6 m) to 10′ (3.0 m) tall and has roots that grow as deep as 4′ (1.2 m).

A. Purple loosestrife/*Lythrum salicaria*

B. Japanese honeysuckle/*Lonicera japonica*

C. Water hyacinth/*Eichhornia crassipes*

D. Cogon grass/*Imperata cylindrica*

41. _____ This herbaceous perennial, which was introduced in the early 1800s, has purple flowers and a clumping growth pattern.

A. Purple loosestrife/*Lythrum salicaria*

B. Japanese honeysuckle/*Lonicera japonica*

C. Water hyacinth/*Eichhornia crassipes*

D. Cogon grass/*Imperata cylindrica*

42. _____ A fast-growing plant that is prone to damage from wind, ice, and snow because of its weak wood.

A. Japanese honeysuckle/*Lonicera japonica*

B. Water hyacinth/*Eichhornia crassipes*

C. Bradford pear trees/*Pyrus calleryana*

D. Cogon grass/*Imperata cylindrica*

43. _____ This vine is aggressive and produces a dense mat that suffocates all other plants beneath it.

A. Purple loosestrife/*Lythrum salicaria*

B. Japanese honeysuckle/*Lonicera japonica*

C. Bradford pear trees/*Pyrus calleryana*

D. Water hyacinth/*Eichhornia crassipes*

44. _____ A perennial plant that is native to the Amazon River basin in South America.

A. Water hyacinth/*Eichhornia crassipes*

B. Purple loosestrife/*Lythrum salicaria*

C. Cogon grass/*Imperata cylindrica*

D. Japanese honeysuckle/*Lonicera japonica*

45. _____ A wildfire is a fire that is burning out of control. Which of the following types of wildfires burns the hottest?

A. Ground fire

B. Crown fire

C. Surface fire

D. Bank fire

46. _____ *True or False?* One of the most detrimental effects of air pollution is acid rain.

47. _____ *True or False?* Global climate change affects precipitation patterns, pest pressures, and acid rain.

48. _____ The main impact of urbanization on forests is _____.

A. noise pollution

B. fragmentation

C. habitat expansion

D. All of the above.

Thinking Critically

1. How does a vascular disease infect a tree?

____

____

2. Why are invasive plant species detrimental to the environment?

____

____

3. For several of the forest pests, the larval stage of the insect's life seems to be the most damaging phase of life. Why do you think this is the case? Why do you think the larval stage may be the most difficult stage in which to kill these insect pests?

____

____

4. How can clearcutting help wildlife?

____

____

5. What does an environmental engineer need to consider when planning to restore a forested area?

____

____

Name ______________________ Date ______________ Period __________

Activity 24A

Fire Safety Messages

Objectives

- Communicate the importance of fire safety with others.
- Identify key themes in fire safety and promote them to others.
- Develop and design a youth-friendly fire safety and awareness message.

Materials

- internet access
- markers, crayons, or colored pencils
- drawing paper

Instructions

Smokey Bear is the symbol of the USDA Forest Service and has been used as a messenger for fire safety since 1944. His message is the longest running public service announcement in the history of the United States and has educated millions with the message "Only You Can Prevent Forest Fires." In some places, Ember the Fox replaced Smokey Bear, but both creatures spread the message about the importance of fire safety. Why is it important to educate all ages about forest safety and fire prevention? How can you help spread this message?

Procedure

1. Research messages on forest safety and what people can do to prevent fires.
2. Design a youth-friendly poster with a key message on fire prevention that displays in pictures and words what needs to be done to prevent forest fires.
3. After creating the poster, present the information and poster to your class.
4. Additional Procedure: bring the completed posters to elementary school classes during fire prevention week and share the topics with others. Use these posters to communicate the importance of fire prevention to forest safety.

Analysis

1. What are five fire prevention tips that all people can follow?
2. If your community has a burn ban, when is it in effect and what is the purpose?
3. What is the cause of most wildfires, and how can the message in your poster help prevent fires?

Notes

Name ______________________________ Date ______________ Period __________

CHAPTER 25 Grasslands and Rangelands

Chapter Review

Carefully study the chapter and then answer the following questions.

1. _____ Why are there few trees in temperate grasslands?
   A. Too hot in summer
   B. Too cold in winter
   C. Insufficient rainfall
   D. All of the above.

2. _____ *True or False?* Rangelands are maintained in part by grazing livestock.

3. _____ Which of the following are grazing animals found in temperate grasslands?
   A. Jackrabbits
   B. Bison
   C. Feral horses
   D. All of the above.

4. _____ Grass consumed for food by grazing animals is _____.
   A. mast
   B. forage
   C. sedge
   D. All of the above.

5. _____ Which of the following are apex predators in temperate grasslands?
   A. Coyotes and vultures
   B. Bison and pronghorns
   C. Coyotes, wolves, and bobcats
   D. Prairie chickens and ruffed grouse

6. _____ Which of the following are scavengers in temperate grasslands?
   A. Coyotes and vultures
   B. Bison and pronghorns
   C. Coyotes, wolves, and bobcats
   D. Prairie chickens and ruffed grouse

7. _____ Which of the following statements is true about flooded grasslands?
   A. They are flooded all year
   B. Commonly used for grazing cattle
   C. They are flooded at least part of the year
   D. Wildfires are common

8. _____ *True or False?* The Everglades is the most well-known flooded grassland in Central America.

9. _____ Which of the following are native to the Everglades?
   A. Pronghorns and alligators
   B. Burmese pythons
   C. Panthers and alligators
   D. All of the above.

10. _____ *True or False?* Montane grasslands are found in mountainous regions.

11. _____ Which of the following cover balds?
   A. Highland forests
   B. Shrubs and trees
   C. Native grasses and trees
   D. Native grasses and shrubs

12. _____ Which of the following are found on the Great Plains?
   A. Tallgrass prairies
   B. Mixed grass prairies
   C. Shortgrass prairies
   D. All of the above.

13. _____ What causes the difference in height of prairie grasses?
   A. Grazing bison and poorer soils
   B. The temperature range and the amount of precipitation
   C. The amount of daylight and humidity
   D. All of the above.

14. _____ Most prairie grasses are _____.
   A. perennials
   B. annuals
   C. bi-annuals
   D. flowering plants

15. _____ Which of the following are forbs commonly found in tallgrass prairies?
   A. Big and little bluestem grass
   B. Sunflowers and purple coneflowers
   C. Cottonwood and red cedar
   D. All of the above.

16. _____ *True or False?* Prairie potholes are abandoned prairie dog burrows.

Name ____

17. _____ Which of the following are found in mixed grass prairies?
   A. Bee balm and black-eyed Susan
   B. Tallgrass and shortgrass
   C. Butterfly milkweed and goldenrod
   D. All of the above.

18. _____ Shortgrass prairies grow where annual precipitation ranges from _____.
   A. 5″ to 10″
   B. 10″ to 12″
   C. 12″ to 24″
   D. 22″ to 30″

19. _____ *True or False?* Wildlife help grasslands remain healthy through seed dispersal, grazing, and nutrient cycling.

20. _____ Wildfires provide many benefits to grasslands, including _____.
   A. killing insect pests
   B. killing weed seeds
   C. burning trees and shrubs
   D. All of the above.

21. _____ Which of the following is used in grassland and rangeland management?
   A. Protecting riparian zones
   B. Mowing several times a season
   C. Planting hybrid grasses
   D. All of the above.

22. _____ Grazing livestock on a relatively small parcel for a short period is _____.
   A. rotational grazing
   B. intensive grazing
   C. alternate grazing
   D. continuous grazing

23. _____ Frequently moving grazing livestock from one parcel to another is _____.
   A. rotational grazing
   B. intensive grazing
   C. alternate grazing
   D. continuous grazing

24. _____ The Bureau of Land Management leases grazing rights for _____.
   A. five years at a time
   B. 10 years at a time
   C. 15 years at a time
   D. 18 years at a time

25. _____ *True or False?* Ranchers pay for grazing rights based on the amount of land needed to sustain one animal unit for one month.

26. _____ *True or False?* Prescribed burns are annual management practice.

27. _____ Which of the following are necessary when conducting a prescribed burn?
   A. Blacklines
   B. Backing and flank fires
   C. Head fires
   D. All of the above.

28. _____ Which of the following are the hottest fires that burn into the wind?
   A. Flank fires
   B. Backing fires
   C. Head fires
   D. Firebreaks

29. _____ Which of the following fires burn with the direction of the wind?
   A. Flank fires
   B. Backing fires
   C. Head fires
   D. Firebreaks

30. _____ Which of the following fires burn in narrow strips?
   A. Flank fires
   B. Backing fires
   C. Head fires
   D. Firebreaks

31. _____ The area that has been burned by a prescribed burn is the _____.
   A. backline
   B. headline
   C. flank line
   D. blackline

32. _____ *True or False?* Grasslands are in danger of becoming deserts because of climate change.

33. _____ *True or False?* Annually, 8.5 million acres are lost to urbanization.

Name ____

Vocabulary Review

Match the following terms with the appropriate definition or description.

A. brush pile
B. food plots
C. bald
D. bird nesting structure
E. forb
F. animal unit
G. riparian zone
H. forest opening

34. _____ Areas of cropland that are left unharvested
35. _____ A mound of tree and shrub limbs
36. _____ Manufactured constructions that provide support for a natural bird nest
37. _____ A hole through the dense vegetation at the edge of the forest
38. _____ Any herbaceous flowering plants
39. _____ Fragile areas between grazing lands and bodies of water
40. _____ Tops of mountains covered in native grasses and shrubs

Thinking Critically

1. Why do prairie grasses withstand wildfires and recover quickly?

____

____

____

2. List four benefits to maintaining healthy rangelands and grasslands.

____

____

____

3. How is carrying capacity used to determine the number of animal units the land can support without damaging the crop?

____

____

____

4. Explain why livestock should have limited interactions in a riparian zone.

____

____

____

5. Identify the precautions that must be taken when conducting a prescribed burn.

6. Explain the benefits of wildlife travel corridors.

7. What are the main threats to grasslands and rangelands?

8. What characteristics of grasslands made them ideal for conversion to farmland? Why? What characteristics of grasslands made or make farming a challenge? Why?

9. How do extremes in temperature affect ranching? Why?

10. Which is more dangerous or damaging to forests, point source or nonpoint source pollution? Why?

Name ______________________ Date ______________ Period __________

Activity 25A

Parts of a Flower

Objectives

- Identify the parts of a flower.
- Explain the function of each part of a flower.
- Explain different ways in which flowers are classified.

Materials

- activity sheet
- colored pencils or fine-point markers
- drawing pencils
- flowering plant stem
- methylene blue
- microscope
- razor blade or craft knife
- slide cover slips
- slides

Instructions

Your instructor will distribute flowers for you to examine. Use your observations to answer the worksheet questions.

Section 1: Examine the Flower

1. Is the flower complete? Why or why not? ______________________________

2. Is the flower perfect? Why or why not? ______________________________

3. Place the flower under the microscope on the lowest power objective. Draw the flower at this magnification. Identify the structures on your drawing.

4. Take one of the stamen and place it under the microscope at medium power, focusing specifically on the anther. Draw the stamen at this magnification. Identify the structures on your drawing.

5. Place the entire pistil under the microscope at medium power. Draw the pistil at this magnification. Identify the structures on your drawing.

6. Carefully remove the stigma from the top of the style and place the stigma topside up beneath the microscope. Draw the stigma at this magnification and identify the structures on your drawing. Do you see anything that would help in plant reproduction? How do you think the magnified structure could be beneficial to pollination and fertilization?

7. Take a small section of the petal and observe it under the highest magnification. Draw the petal section at this magnification. Identify the structures on your drawing.

Name ______________________ Date ____________ Period ________

Activity 25B

Parts of a Stem

Objectives

- Identify the parts of a plant stem.
- Explain the function of each part of a plant stem.

Materials

- activity sheet
- colored pencils or fine-point markers (optional)
- drawing pencils
- flowering plant stem (one per student)
- methyl blue
- microscope (one per pair of students)
- razor blade or craft knife (one per student)
- slide cover slips
- slides

Instructions

Your instructor will distribute stems for you to examine. Use your observations to answer the worksheet questions.

Plant Stems

Plants may have herbaceous stems or woody stems. Plants that do not produce wood have herbaceous stems, which are flexible and less rigorous than woody stems. Plants that produce wood include trees and most shrubs. Herbaceous stems continually grow upward and have their xylem and phloem in a bundle. Woody stems grow upward and outward. The xylem and phloem in woody stems grow in large rings around the stem. Hence, a tree's annual rings.

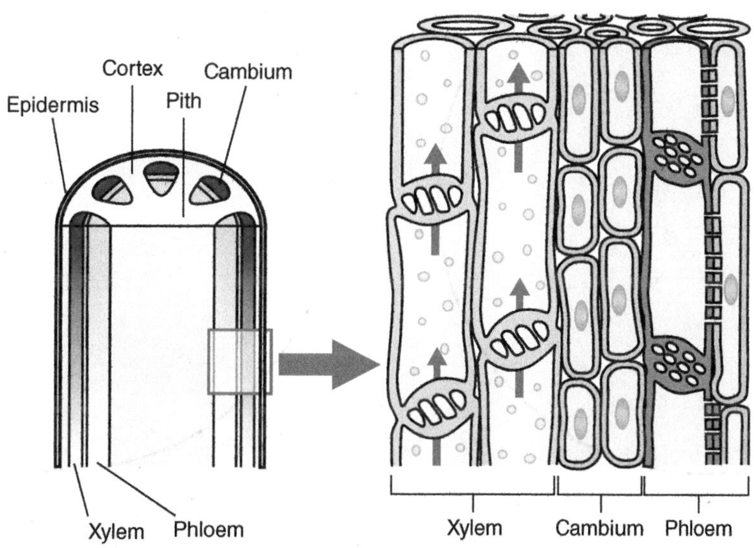

Goodheart-Willcox Publisher

Examine the Stem

1. Carefully cut a section of the stem. Place the section under the microscope at low power. Draw the stem section at this magnification. Identify the structures on your drawing.

2. Carefully cut a cross-sectional piece (a thin circular slice) of the stem. Place the slice cut side up on the stage. Draw what you see at this magnification. Identify the structures on your drawing. Are the structures organized or scattered through the stem piece?

____

3. Cut a very thin section of the stem from the larger piece. Place this piece on a slide and add a drop of methyl blue to the slide. Use a cover slip and place the slide under the microscope at medium power. Draw the stem section at this magnification. Identify the structures on your drawing.

4. Increase the power to the highest objective. Draw the stem section at this magnification. Identify the structures on your drawing.

Name ______________________ Date ____________ Period ________

Activity 25C

Bureau of Land Management Wild Horses and Burros Control Method Debate

Objectives

- Identify issues with wild animal control methods.
- Improve research and data collection skills.

Materials

- internet access
- activity sheet

Instructions

The Bureau of Land Management has often conducted controversial methods of reducing wild horse and burro populations to maintain grazing land for other species. Visit the BLM website at blm.gov and read about past and current methods to control the population of wild horses and burros in the United States. Split up into groups and hold a debate on the ethics of these methods including but not limited to:

- Sales of horses for slaughter
- Capture and auction of wild horses and burros
- Capture and breaking of wild horses and burros
- Fertility vaccinations of mares
- Sterilization of breeding stallions
- Predator introduction to reduce population size

1. What would happen to wild horses and burros if the population were left unchecked?

   ____

2. Why is this issue important to grasslands and rangelands?

   ____

3. Which method do you feel is the best option?

   ____

4. Which method do you feel is the worst option?

   ____

Notes

Name ______________________ Date ____________ Period ________

Activity 25D

Beef Cattle on the Range

Objectives

- Identify issues with grazing on public and private lands.
- Improve research and data collection skills.

Materials

- activity sheet

Instructions

You will be examining grazing issues on public and private rangelands. Research topics by speaking with local beef cattle producers, reading through beef cattle magazines, or searching online. Once your teacher has approved the topic, research the subject and complete this worksheet.

1. Describe the issue you have selected.

____
____
____
____
____
____

2. Why is this issue important to beef cattle management?

____
____
____
____
____
____

3. List at least five facts about the issue that you have found in your research.

____
____
____
____
____
____
____

4. What are the options for resolving the issue?

5. What do you think should be done concerning this issue?

6. To extend the activity, prepare an argument for debate.

Name ______________________________ Date ______________ Period __________

CHAPTER 26 Outdoor Recreation

Chapter Review

Carefully study the chapter and then answer the following questions.

1. _____ *True or False?* Outdoor recreation can provide physical and mental health benefits.

2. _____ Which of the following is used by birders to help find birds?
   A. Feeding habits
   B. Flight patterns
   C. Audio recordings
   D. All of the above.

3. _____ *True or False?* Walking trails for extended distances and camping overnight is hiking.

4. _____ *True or False?* Walking trails in the woods for periods less than a day is backpacking.

5. _____ *True or False?* It is often illegal to cut flowers or trees in wilderness areas.

6. _____ *True or False?* Campers use a variety of shelter, including RVs, sleeping bags, and tents.

7. _____ Campers should pitch tents or park RVs in designated camping areas to _____.
   A. avoid disturbing animals
   B. protect plant species
   C. get the best view
   D. be near the main roads

8. _____ *True or False?* A fire ring of steel or stone should be used when building a campfire.

9. _____ *True or False?* Mountain bikes should be maintained in good working order before taking to mountain trails.

10. _____ Which of the following apply to trail running?
    A. Wearing appropriate clothing
    B. Wearing specially designed shoes
    C. Leaving your itinerary with a responsible party
    D. All of the above.

11. _____ In order to fish legally, fishers must purchase a _____.
    A. life jacket
    B. fishing license
    C. map
    D. All of the above.

12. _____ *True or False?* Hunting is used as a means of wildlife management.

13. _____ *True or False?* Hunters should wear bright orange to be less visible to animals.

14. _____ *True or False?* Horses should be kept on trails to avoid causing damage to natural areas.

15. _____ An outdoor activity that requires a GPS unit to find hidden caches is _____.
    A. camper's hide and seek
    B. find the hidden object
    C. geocaching
    D. All of the above.

16. _____ *True or False?* ATVs must be kept out of streams to prevent damage to the surrounding area and contaminating the water.

17. _____ *True or False?* Alcohol is the leading cause of boating fatalities in the United States.

18. _____ *True or False?* Snowmobile riders are prohibited from riding on private property in all states.

19. _____ *True or False?* Frostbite should be treated immediately with a heating device.

20. _____ *True or False?* When treating someone for hypothermia, you should warm the body as quickly as possible.

21. _____ Which of the following is a symptom of heat stroke?
    A. Muscle weakness or cramps
    B. Headache and confusion
    C. Rapid and shallow breathing
    D. All of the above.

22. _____ *True or False?* If stream water appears clear and is moving rapidly, it should be safe to drink.

23. _____ *True or False?* Sustainability in outdoor recreation means that those involved in any outdoor activity should leave minimal, if not any, trace.

24. _____ *True or False?* It is the responsibility of every outdoor enthusiast to take whatever they bring into the wilderness out of the wilderness when they leave.

25. _____ *True or False?* The largest segment of outdoor recreation is camping with recreational vehicles.

26. _____ The outdoor recreation industry in the United States contributes how much to the US GDP?
    A. $262 billion
    B. $462 billion
    C. $626 billion
    D. $862 billion

Name ______________________________

Thinking Critically

1. Why should all outdoor enthusiasts leave a copy of their itinerary with someone responsible before they travel into the wilderness?

2. List five supply items that anyone participating in outdoor recreational activities should have on hand.

3. What four methods can be used to purify drinking water?

4. How do you show good stewardship of natural resources in your home?

5. What is the history of fair chase rules? Why are fair chase rules enacted? What would be an example of an unfair chase?

6. Look up hunting seasons for game species in your state. Analyze the species of game. What are the specific biological reasons for limiting the hunting season for these species?

Notes

Name ______________________________ Date ______________ Period __________

Activity 26A

Safety Pamphlet

Objectives

- Use research and writing skills to create a safety pamphlet.
- Design and produce a safety pamphlet.

Materials

- online access
- printer and paper

Instructions

For this activity, you will be creating an informational pamphlet about safety for an outdoor recreation activity.

1. Select a specific type of outdoor recreation, such as turkey hunting, cross-country skiing, snowmobiling, or mountain biking.
2. What is your activity?

   ____

3. Once you have your teacher's approval, research your topic and gather data for your pamphlet.
4. Describe your activity.

   ____
   ____
   ____
   ____
   ____
   ____

5. What types of safety gear are required for the activity?

   ____
   ____
   ____
   ____
   ____

6. Does the activity require a permit or license?

   ____

7. If so, what are the restrictions for the permit (daily, annual, seasonal, local, national, etc.)?

8. Are there age restrictions for the activity?

9. Are there classes available? If so, where?

10. Where can this activity be performed?

11. List additional tips and guidelines for safely participating in the activity.

12. Using the information you have gathered, create a trifold safety pamphlet for your outdoor activity. Include images to enhance your pamphlet and its message.

Name ______________________________ Date ______________ Period __________

CHAPTER 27 National Protected Areas

Chapter Review

Carefully study the chapter and then answer the following questions.

1. _____ Which of the following manage natural protected areas in the United States?
   A. Bureau of Land Management and National Oceanic and Atmospheric Administration
   B. National Park Service and Department of Defense
   C. US Forest Service and US Fish and Wildlife Service
   D. All of the above.

2. _____ Which of the following began in 1891 when it was signed by President Benjamin Harrison?
   A. Agricultural Marketing Act
   B. Forest Reserve Act
   C. Antiquities Act
   D. Natural Trails System Act

3. _____ Which of the following Acts was used to remove marginal lands from cultivation and create our national grasslands?
   A. Agricultural Marketing Act
   B. Forest Reserve Act
   C. Antiquities Act
   D. Natural Trails System Act

4. _____ *True or False?* National grasslands are open to grazing and mining, as well as hunting, fishing, hiking, and horseback riding.

5. _____ Which of the following was the first official wildlife refuge?
   A. The Everglades
   B. Pelican Island
   C. Yosemite
   D. Midewin

6. _____ Which of the following was established as a national monument by President Theodore Roosevelt?
   A. Baldwin Tower
   B. Pelican Tower
   C. Devils Tower
   D. Rocky Tower

7. _____ Which of the following is prohibited in national parks but allowed in national forests through the multiple use mandate?

A. Logging

B. Mining

C. Hunting

D. All of the above.

8. _____ The National Parks Service has an annual budget of nearly $3 billion to maintain _____.

A. 2.5 million acres of national park land

B. 15 million acres of national park land

C. 24 million acres of national park land

D. 35 million acres of national park land

9. _____ *True or False?* To qualify as a national wild and scenic riverway, the river must not be obstructed by damming, diversion, or channeling.

10. _____ *True or False?* The country's national lakeshores are located around the Great Lakes.

11. _____ How many national seashores have been established on the Pacific, Atlantic, and Gulf Coasts?

A. 3

B. 6

C. 10

D. 12

12. _____ Which of the following statements about national scenic trails is accurate?

A. There are 16 national scenic trails

B. They are short-distance hiking paths

C. They are authorized under the National Trails System Act of 1968

D. They are designed primarily for small motorized vehicles

13. _____ *True or False?* A national parkway is a scenic camping area on national land.

14. _____ Which of the following have national memorials?

A. Martin Luther King, Jr.

B. Flight 93 (hijacked on 9/11)

C. Thaddeus Kosciuszko

D. All of the above.

15. _____ Which of the following maintains the National Register of Historic Places?

A. FWS

B. NOAA

C. NPS

D. BLM

16. _____ Which of the following is true about national wilderness areas?

A. They are protected and maintained in their natural state

B. They cover 210 million acres

C. They are managed by the Department of Defense

D. All of the above.

Name ______________________________

17. _____ *True or False?* Motorized vehicles and mining are allowed in national conservation areas.

18. _____ *True or False?* Private lands are owned by an individual or corporation.

19. _____ *True or False?* Public lands are owned by the US Federal Government.

20. _____ *True or False?* Part of the motivation for creating national parks was to avoid commercialism around or near our natural wonders.

21. _____ The total acreage of state-supported conservation areas is almost _____.
    A. 10 million acres
    B. 19 million acres
    C. 25 million acres
    D. 52 million acres

Vocabulary Review

Match the natural protected areas to the corresponding definition or description.

A. national conservation areas
B. national estuarine reserves
C. national marine sanctuaries
D. national monument
E. national park
F. national preserve
G. national protected areas
H. national recreation areas
I. national wildlife refuges

1. _____ Sites or features that the US Federal Government has determined to be significant and worthy of protection

2. _____ The system of public lands and waters that have been set aside to conserve the nation's plants and wildlife, including fish

3. _____ A protected area of land or water that has been given permanent protection by Congress or by the US President through the Antiquities Act

4. _____ An area that is set aside by the federal government for the preservation of the natural environment

5. _____ An area with characteristics associated with national parks, but in which hunting, trapping, fishing, and mining are permitted

6. _____ Protected waters that include habitats such as rocky reefs, kelp forests, and underwater archeological sites

7. _____ Research areas where a freshwater river merges with the ocean and mixes with saltwater

8. _____ Protected areas that are centered on large reservoirs and emphasize water-oriented recreation

9. _____ Areas managed by the Bureau of Land Management on which most motorized vehicles are prohibited but activities such as hunting and fishing are permitted in some locations

Thinking Critically

1. What is a national historic trail?

2. What qualifications must land meet to be designated as a national wilderness area?

3. If a wetland were going to be drained for construction of a shopping mall, how would you argue that the wetland provides more value than the shopping mall?

4. How would you argue to a farmer that he or she should convert a portion of his or her productive farmland and convert it into a riparian zone? What are the benefits for wildlife, the environment, and even the farmer?

5. Which event in the history of natural resources conservation would you consider to be the most important? Why?

Name ______________________ Date ____________ Period ________

CHAPTER 28 Surveying and Cartography

Chapter Review

Carefully study the chapter and then answer the following questions.

1. _____ *True or False?* Land surveying is used to determine boundaries.
2. _____ A total station contains which of the following?
   A. An electronic transit
   B. Distance measuring
   C. Computer for calculations
   D. All of the above.
3. _____ How many GPS satellites are controlled by the US Air Force?
   A. 4
   B. 12
   C. 26
   D. 31
4. _____ *True or False?* Cartography is the science of reading maps.

For questions 5–14, match the map components to the corresponding description.

A. compass rose
B. relief
C. contour interval
D. contour lines
E. legend
F. elevation
G. title
H. gradient
I. index lines
J. scale

5. _____ Indicates the area the map represents
6. _____ The lines that loop and wave around the map
7. _____ Indicates which directions are indicated by various markers
8. _____ Every fifth line on which the exact elevation is indicated
9. _____ Explains the scientific methods used to create the map and the symbols used
10. _____ The vertical distance between the lines
11. _____ A given point measured by its height above sea level
12. _____ Provides a ratio of map distance to actual distance
13. _____ The difference in elevation found in an area
14. _____ The steepness of an area

15. _____ *True or False?* Topographic maps show features using contour lines to show changes in the landscape.

16. _____ An instrument used for navigation and orientation relative to the cardinal directions is a _____.
   A. magnometer
   B. compass
   C. theodolite
   D. transit

17. _____ The number of degrees between an object and north on your compass is the compass _____.
   A. bearing
   B. orientation
   C. declination
   D. scale

For questions 18–27, identify the parts of a baseplate compass.

18. _____ azimuth circle
19. _____ base plate
20. _____ declination scale
21. _____ direction of travel arrow
22. _____ index line
23. _____ magnetized needle
24. _____ orienting arrow
25. _____ orienting lines
26. _____ rotating housing
27. _____ scales/rulers

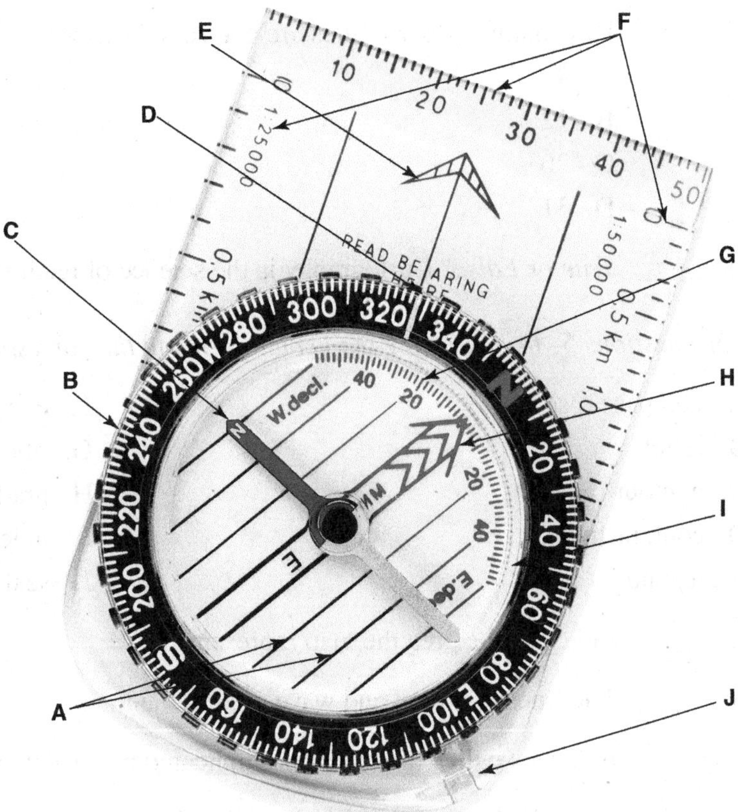

RG-vc/Shutterstock.com

Name ______________________________

Thinking Critically

1. What are the three major components of GPS?

2. What is the Landsat Program?

3. Identify three uses for GIS.

4. How are drones used in natural resources applications?

5. What is the difference between magnetic north and true north?

6. From a land capability classification perspective, why is some land more valuable than others for agricultural purposes?

7. What does your local county Soil and Water Conservation District do to educate farmers and the public about soil and water conservation? What opportunities exist for youth to be involved in these programs?

Notes

Name ______________________ Date ______________ Period __________

Activity 28A

Measuring Worksheet

Objectives

- Accurately measure various items.
- Properly add, subtract, multiply, and divide measurements.
- Identify instances in various natural resources jobs when accurate measurements and calculations would be necessary.

Materials

- activity sheet
- items (real or imitation) for measuring, such as leaves, seeds, stems, flowers, mushrooms, and insects
- tape measure or ruler

Instructions

Pair up with a fellow student. Your teacher will provide several labeled groups (A–F) of materials from which you and your partner are to select items for measuring, or you will spend time collecting items before performing this activity.

1. With your partner, collect one item from each of the labeled groups of materials.
2. Individually, measure the length of each item to the nearest 1/16 of an inch. Write the description of each item and its measurement in the space provided.

   A. Item: ______________________ Measurement: ______________________

   B. Item: ______________________ Measurement: ______________________

   C. Item: ______________________ Measurement: ______________________

   D. Item: ______________________ Measurement: ______________________

   E. Item: ______________________ Measurement: ______________________

   F. Item: ______________________ Measurement: ______________________
3. Compare your initial measurements to your partner's. If your measurements differ for any of the items, measure the materials again until you come to a consensus. Convert your measurements to sixteenths.

4. Perform the following calculations and write your answers in the spaces provided. Include your equation. (For example: A + B = 1 + 2 = 3)

   A. What is the length of A + B?

   ____

   ____

   B. What is the length of C + D?

   ____

   ____

   C. What is the length of E + F?

   ____

   ____

   D. What is the length of all six items (A through F) if they are placed end to end?

   ____

   ____

5. Perform the following calculations and write your answers in the spaces provided. Include your equation. (For example: A – B = 2 – 1 = 1)

   A. Subtract the length of your shortest item from your longest item.

   ____

   ____

   B. Add the lengths of A through E, then subtract the length of F.

   ____

   ____

6. Perform the following multiplication calculations and write your answers in the spaces provided. Include your equations.

   A. What would be the total length of six pieces of item A laid end to end?

   ____

   ____

   B. What would be the total length of four pieces of item B and five pieces of item D laid end to end? Include your equation.

   ____

   ____

Name ____

7. Perform the following division calculations and write your answers in the spaces provided. Include your equation.

   A. Divide the length of item G into three equal pieces. How long is each piece?

   ____

   ____

   B. Divide the length of item E into 1/2" pieces. How many 1/2" pieces are there? Is there an equal number of pieces?

   ____

   ____

8. Was it difficult to obtain accurate measurements on some items? If so, explain wh

   ____

9. Was it more difficult to add, subtract, multiply, or divide the measurements? Explain your answer.

   ____

   ____

   ____

10. Give a real-world example of a time when someone might need to add measurements.

    ____

    ____

    ____

11. Give a real-world example of a time when someone might need to subtract measurements.

    ____

    ____

    ____

12. Give a real-world example of a time when someone might need to multiply measurements.

    ____

    ____

    ____

13. Give a real-world example of a time when someone might need to divide measurements.

    ____

    ____

    ____

Notes